MACMILLAN/McGRAW-HILL

Math

Daily Practice Workbook with Summer Skills Refresher

Grade 1

The McGraw·Hill Companies

Published by Macmillan/McGraw-Hill, of McGraw-Hill Education, a division of The McGraw-Hill Companies, Inc., Two Penn Plaza, New York, New York 10121.

Printed in the United States of America

6 7 8 9 079 08 07 06

Contents

Daily Practice

Summer Skills Refresher

Name ______________________________

Sort with Venn Diagrams

Sort the cubes. Draw the cubes on the Venn Diagram.

1. Use 2 , 3 , 2 , 1

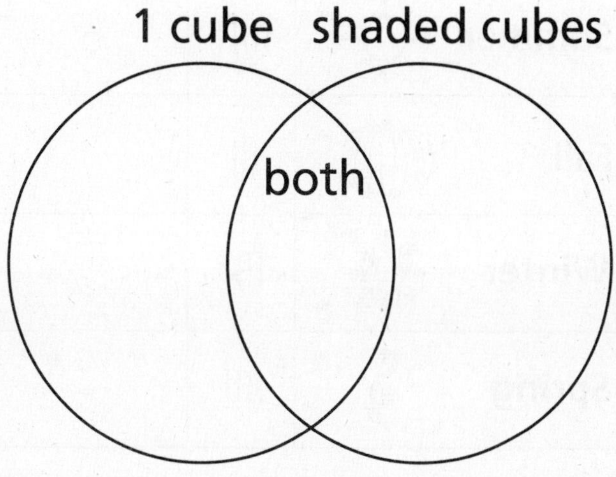

2. Use 2 , 1 , 2 , 1 , 1

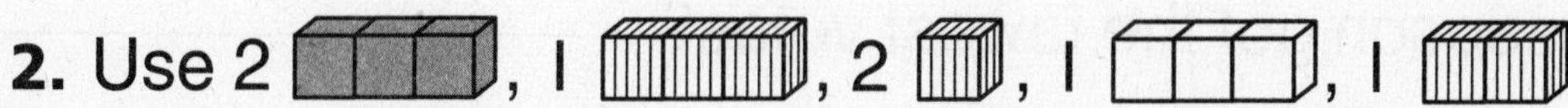

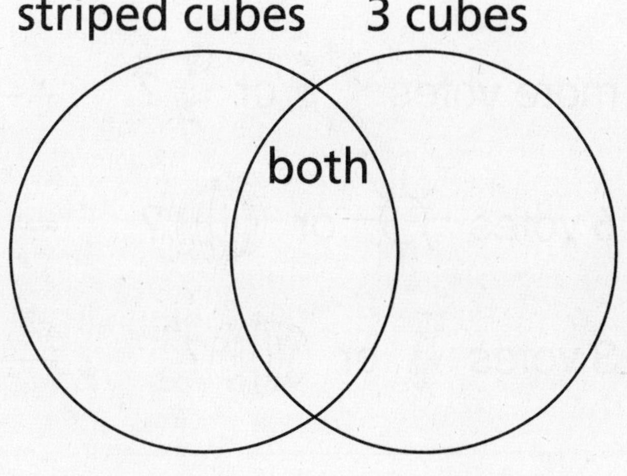

Name ________________________________

Tally Charts

Count the tally marks. Write each total.

My Favorite Season		
Season	**Tally**	**Total**
Summer	𝍸I	6
Fall	III	
Winter	𝍸	
Spring	IIII	

1. Which season got the most votes? ____________

2. Which season got the fewest votes? ____________

3. How many chose ? ____________

4. Which got more votes or ? ____________

5. Which got 6 votes or ? ____________

6. Which got 3 votes or ? ____________

Name ______________________________

Picture Graphs

Use the graph to answer the questions.

Favorite Fruit	
apple	
banana	
strawberry	

1. Do more people like  or 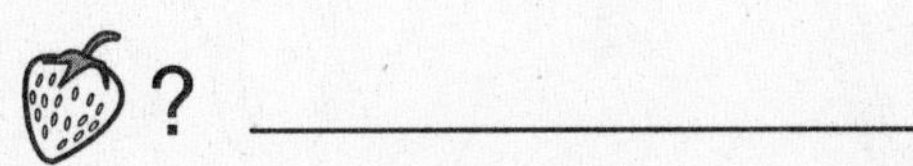 ? ______________

2. Which fruit has 3 votes? ______________

3. How many people like 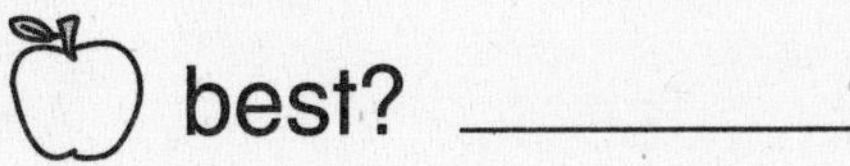 best? ________

4. Which fruit has more than 5 votes? ______________

5. There are how many more votes for than for ? ________

6. There are how many more votes for than for ? ________

Name ______________________________

Problem Solving Skill: Reading for Math

Compare and Contrast

 have 4 legs.

 have 2 legs.

 have 0 legs.

1. Which animal has more than 2 legs?
Circle it.

2. How many are in the picture? ______________

3. Do more animals in the picture have 0 legs or 2 legs? ______________

4. How many more are there than ? ______________

Name ____________________

Numbers to 10

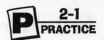

2-1 PRACTICE

Count how many .

1.

2.

3.

4.

5.

6.

7.

8.

Name ______________________________

Explore Zero

Count the sheep. Write the number.

1.  3

2. 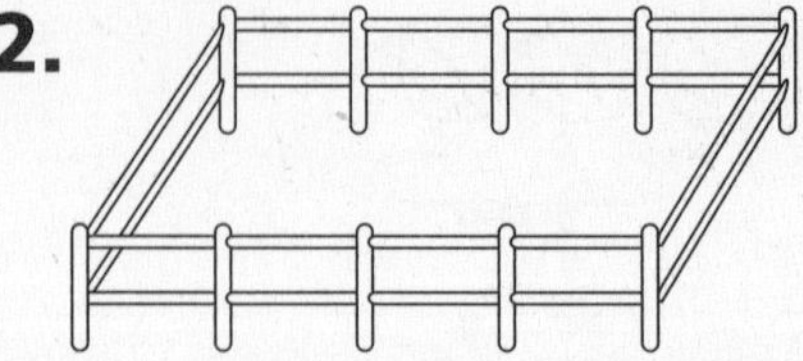

3. 

4. 

5. 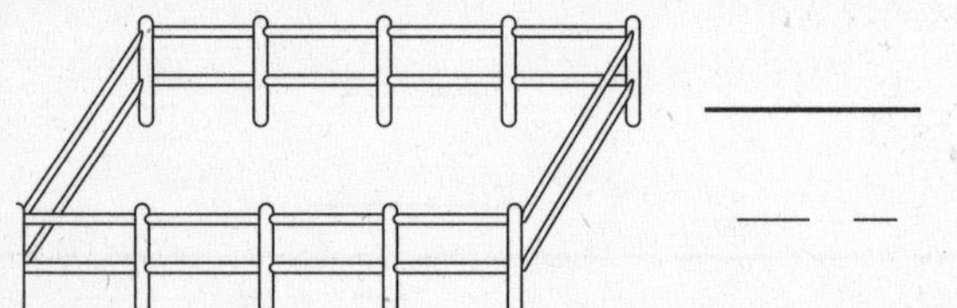

6. 

Use the picture. Write how many.

7. 

8. 

9. 

Name ___

Numbers to 20

Use ten-frame and ○.

Color. Write how many tens and ones.

1. 12 twelve

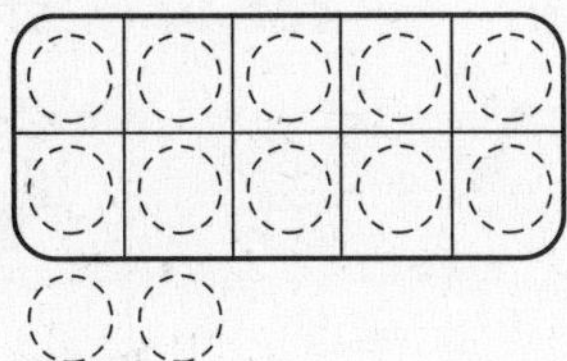

1 ten 2 ones

2. 16 sixteen

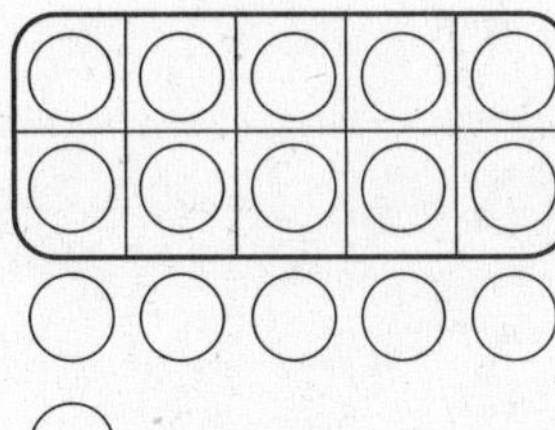

___ ten ___ ones

3. 15 fifteen

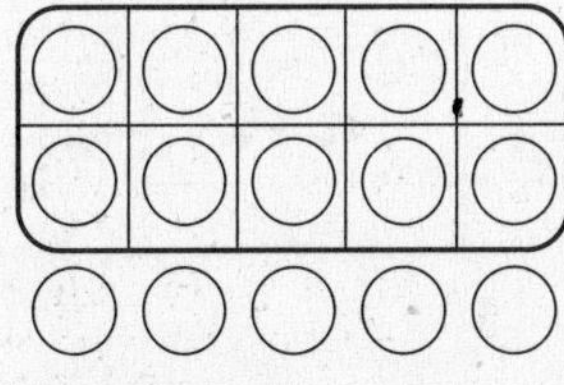

___ ten ___ ones

4. 11 eleven

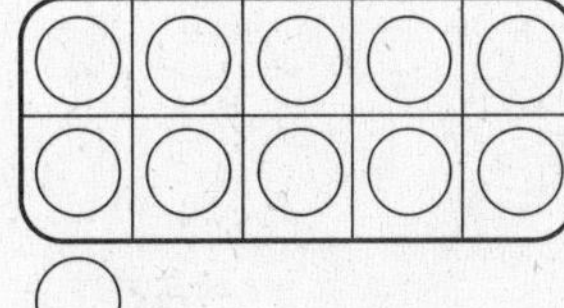

___ ten ___ one

5. 20 twenty

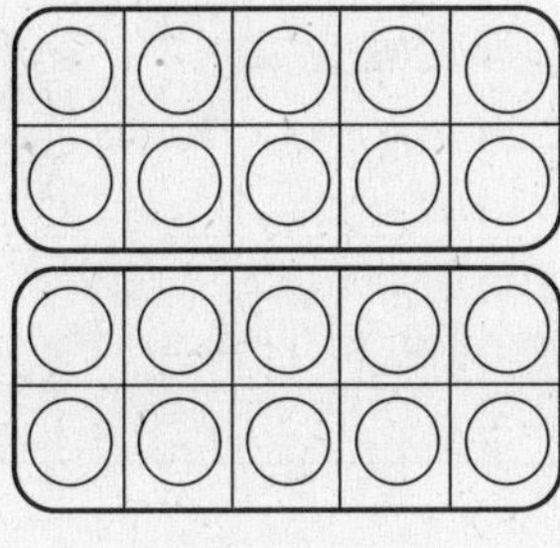

___ tens ___ ones

6. 18 eighteen

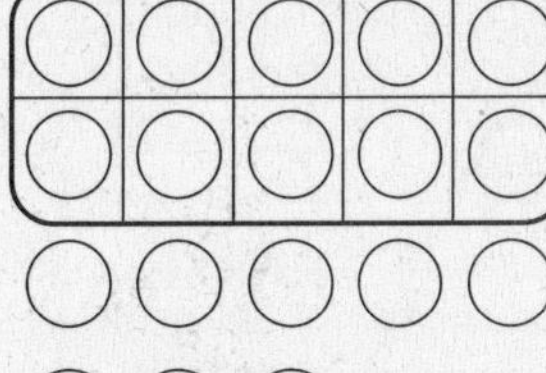

___ ten ___ ones

Name ____________________

Problem Solving: Strategy

Make a Table

Use the picture. Make a table to solve.

1.

Farm Animals			
Animal	horse	cat	sheep
How Many?			

Which animal do you see the most? __________

2.

Farm Animals			
Animal	dog	pig	cow
How Many?			

How many more pigs are there than cows? __________

3.

Farm Animals			
Animal	sheep	horse	duck
How Many?			

Are there more ducks or more sheep?

more __________

Name ____________________

Numbers to 31

3-1 PRACTICE

Write how many tens and ones. Then write the number.

1.

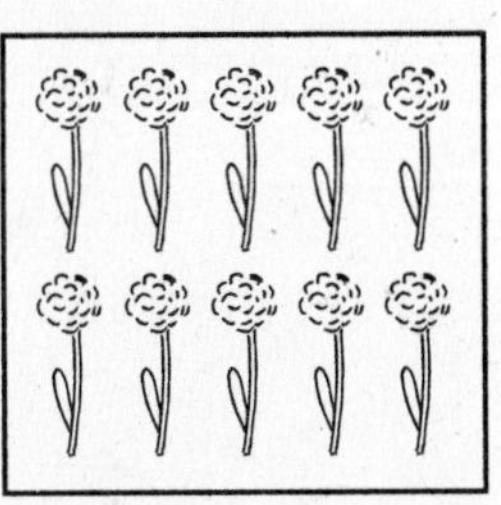

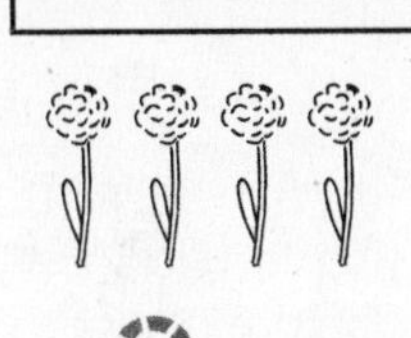

24

twenty-four

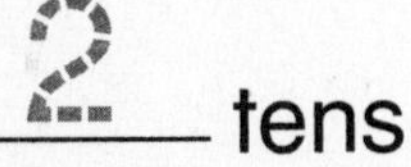

2 tens 4 ones

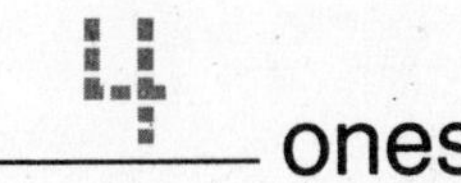

2.

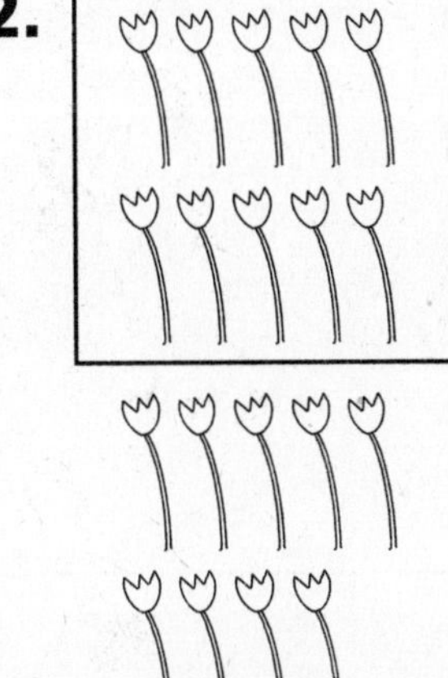

______

nineteen

______ ten ______ ones

3.

______

twenty-eight

______ tens ______ ones

4.

______

thirty-one

______ tens ______ one

Name ______________________________

Compare Numbers

Count how many. Write the number.
Circle the group that shows fewer.

1. 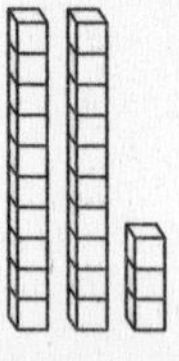

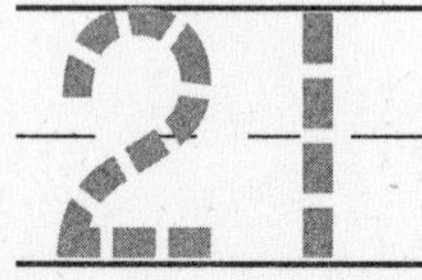

2. 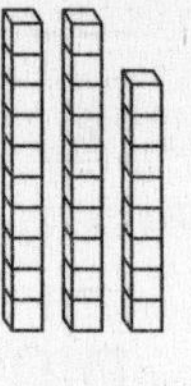

______ ______

3. 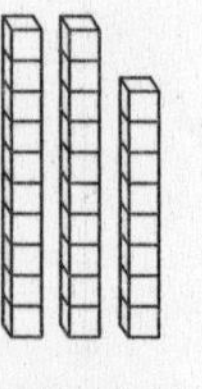

______ ______

4. 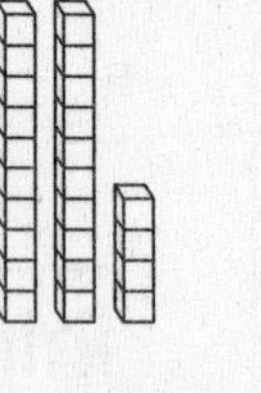 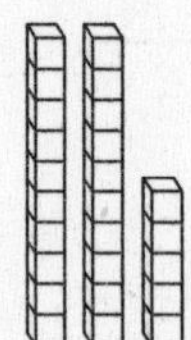

______ ______

Use with Grade 1, Chapter 3, Lesson 2, pages 33–34.

Name ______________________

Greater Than and Less Than

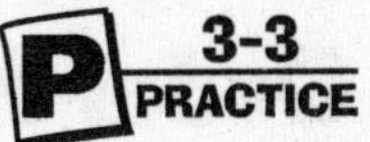

Compare.
Circle greater or less.

Use ▢ for help.

Compare the tens first. The number with more tens is greater.

1. 25 is ________ than 15.
   greater   less

2. 18 is ________ than 31.
   greater   less

3. 22 is ________ than 19.
   greater   less

4. 30 is ________ than 21.
   greater   less

5. 18 is ________ than 28.
   greater   less

6. 26 is ________ than 12.
   greater   less

7. 11 is ________ than 29.
   greater   less

8. 13 is ________ than 20.
   greater   less

9. 27 is ________ than 17.
   greater   less

10. 23 is ________ than 10.
   greater   less

Name ______________________________

Order Numbers to 31

P 3-4 PRACTICE

Write each missing number.

You can use the number line.

20 21 22 23 24 25 26 27 28 29 30 31

---

1. 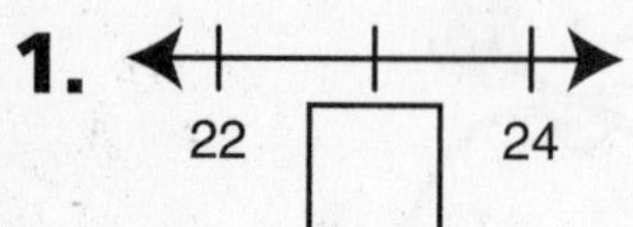

22   23   24

2. 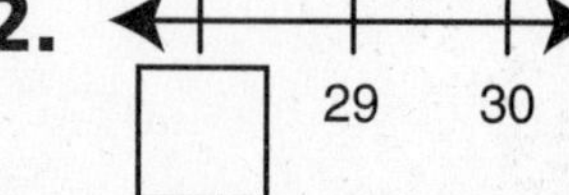

____   29   30

3. 25 26 ☐

25   26   ____

4. 20 ☐ 22

20   ____   22

5. 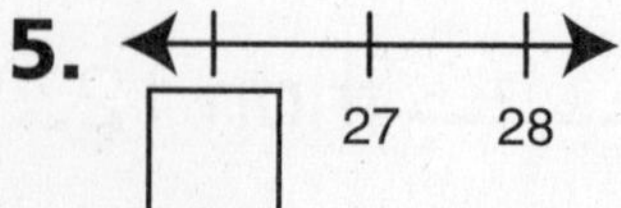

____   27   28

6. 29 ☐ 31

29   ____   31

7. 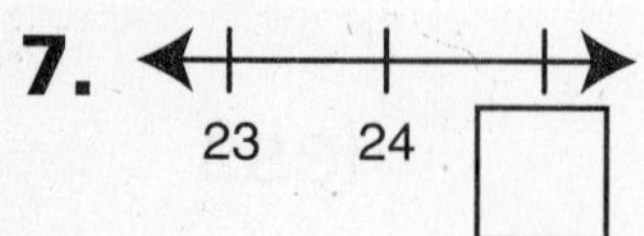

23   24   ____

8. 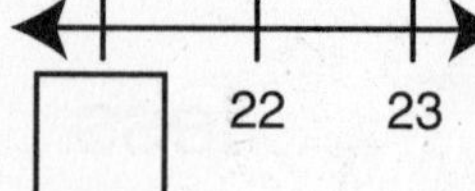

____   22   23

9.

30		28			25	

30 ____ 28 ____ ____ 25 ____

Name ____________________

Ordinal Numbers

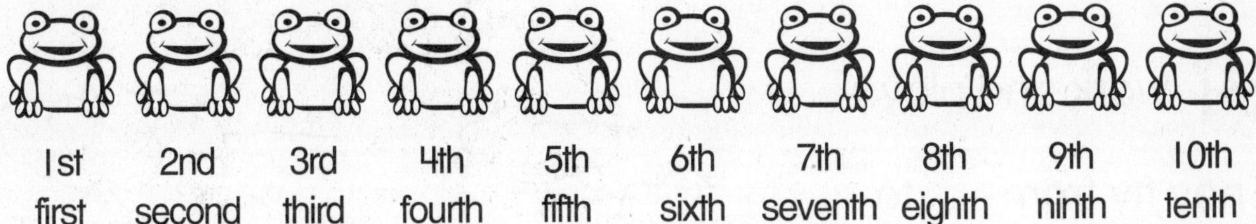

1st	2nd	3rd	4th	5th	6th	7th	8th	9th	10th
first	second	third	fourth	fifth	sixth	seventh	eighth	ninth	tenth

Find first. Then circle the frogs to show order.

1. Circle the eighth frog.

2. Circle the third frog.

3. Circle the ninth frog.

4. Circle the fifth frog.

Name ______________________________

Problem Solving Skill: Reading for Math

Use Illustrations

work and play.

They fly from  to .

Then they go home.

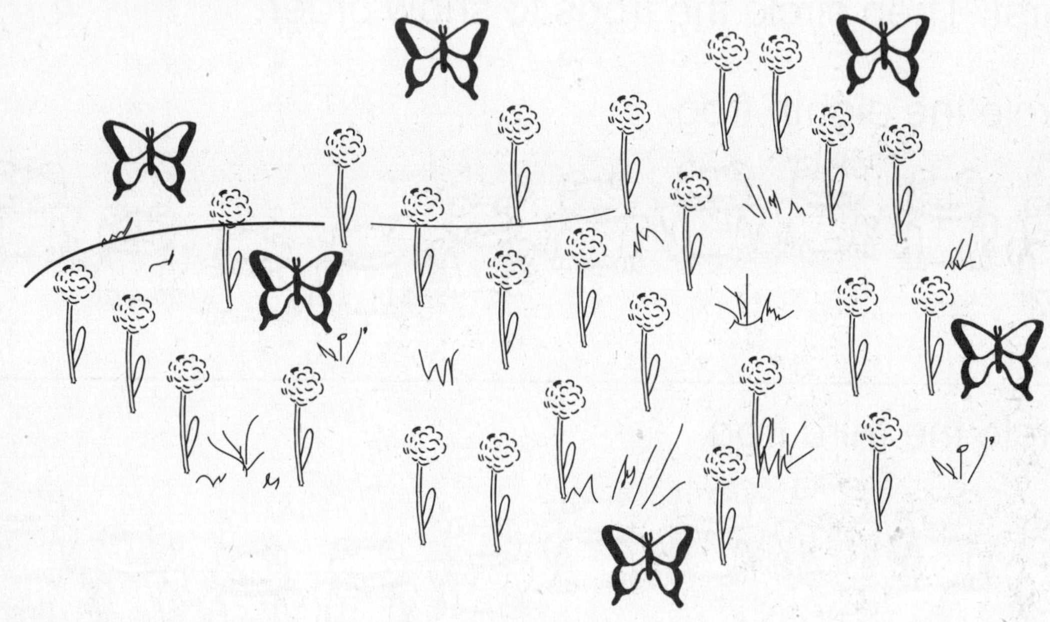

1. Count the . How many  are there? _____ 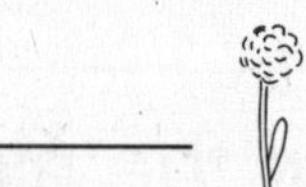

2. Show the number of as tens and ones.

   _____ tens _____ ones

3. How many are there? _____

4. Draw 1 for each .

Name ____________________

Number Stories

Use ◯ in the space below to help show number stories.

Find how many in all.

1. Show 5.
   Show 1 more.
   How many in all?   ____

2. Show 4.
   Show 4 more.
   How many in all? ____

3. Show 3.
   Show 4 more.
   How many in all? ____

Find how many are left.

4. Show 6.
   Take 4 away.
   How many are left? ____

5. Show 4.
   Take 1 away.
   How many are left? ____

6. Show 7.
   Take 2 away.
   How many are left? ____

Name ____________________

Addition Sentences • Algebra

Write the addition sentence.

1.

3 and 1 is 4.

3 (+) 1 (=) 4

2.

2 and 2 is 4.

___ ◯ ___ ◯ ___

3.

3 and 2 is 5.

___ ◯ ___ ◯ ___

4.

4 and 1 is 5.

___ ◯ ___ ◯ ___

5.

4 and 2 is 6.

___ ◯ ___ ◯ ___

6.

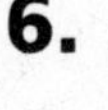

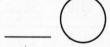

3 and 3 is 6.

___ ◯ ___ ◯ ___

Use with Grade 1, Chapter 4, Lesson 2, pages 53–54.

Name ____________________

Subtraction Sentences • Algebra

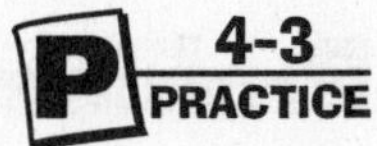

Write the subtraction sentence.

1.

3 take away 1 is 2

3 ◯− 1 ◯= 2

2.

4 take away 2 is 2

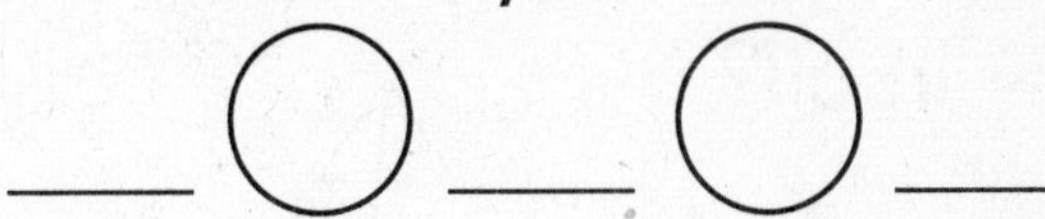

____ ◯ ____ ◯ ____

3.

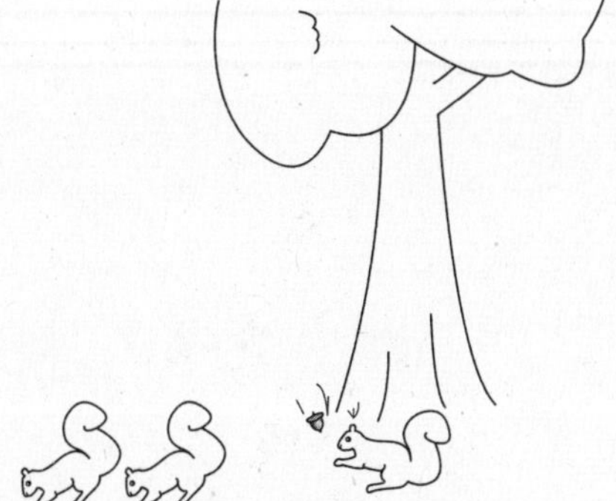

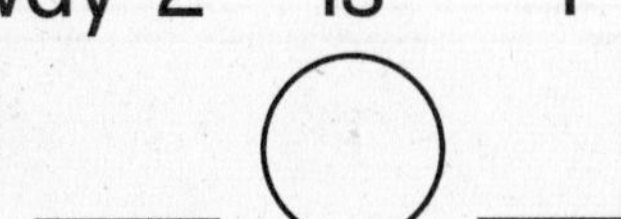

3 take away 2 is 1

____ ◯ ____ ◯ ____

4.

2 take away 1 is 1

____ ◯ ____ ◯ ____

5.

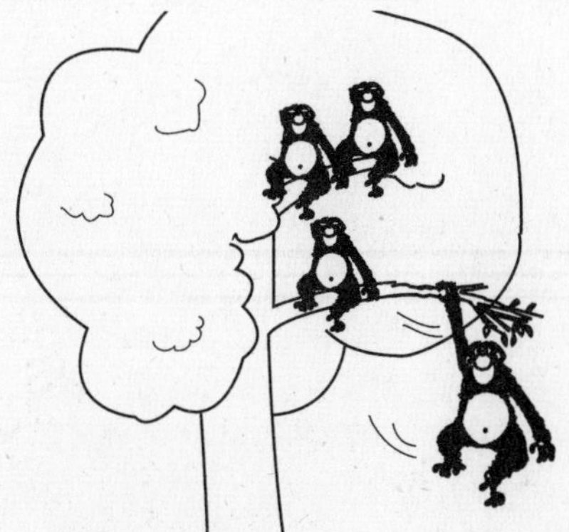

4 take away 1 is 3

____ ◯ ____ ◯ ____

6.

6 take away 3 is 3

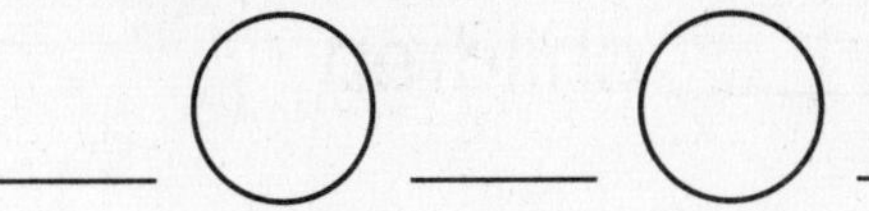

____ ◯ ____ ◯ ____

Name ____________________

Problem Solving: Strategy

Draw a Picture

Draw a picture to solve.

1. Betsy makes 2 kites.
Luís makes 1 kite.
How many kites are made in all?

 kites

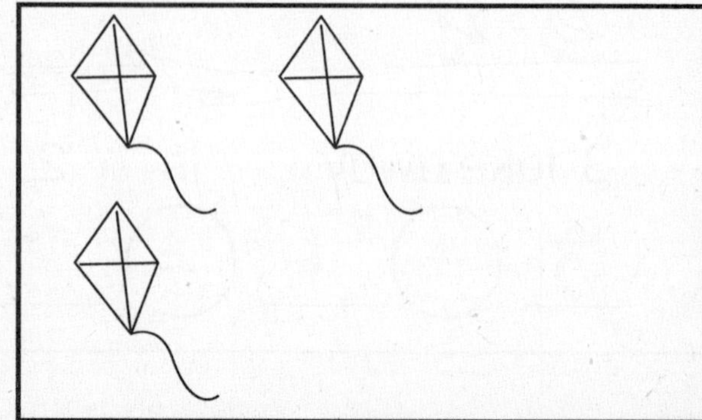

2. There are 5 rolls of string.
4 rolls are used.
How many rolls of string are left?

______ rolls of string 

3. Blake finds 4 crayons.
Kim finds 5 more.
How many crayons do they have in all?

______ crayons 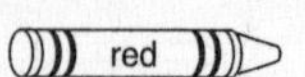

4. 6 children play ball.
3 children leave.
How many children are left?

______ children 

Use with Grade 1, Chapter 4, Lesson 4, pages 57–58.

Name ______________________________

Ways to Make 4, 5, and 6

Put ○ in two groups to make 4, 5, and 6.
Write the numbers.

●	plus	○	equals	sum
____	+	____	=	4
____	+	____	=	4
____	+	____	=	4
____	+	____	=	5
____	+	____	=	5
____	+	____	=	5
____	+	____	=	5
____	+	____	=	6
____	+	____	=	6
____	+	____	=	6
____	+	____	=	6
____	+	____	=	6

Name ________________________________

Ways to Make 7, 8, and 9

Put ◯ in two groups to make 7, 8, and 9.
Write the numbers.

●	plus	◯	equals	sum
____	+	____	=	7
____	+	____	=	7
____	+	____	=	7
____	+	____	=	7
____	+	____	=	8
____	+	____	=	8
____	+	____	=	8
____	+	____	=	8
____	+	____	=	9
____	+	____	=	9
____	+	____	=	9
____	+	____	=	9
____	+	____	=	9
____	+	____	=	9
____	+	____	=	9
____	+	____	=	9

Name ____________________

Ways to Make 10

Use ◯ and ▭ to make 10.

Draw the ◯. Write the addition sentence.

1. 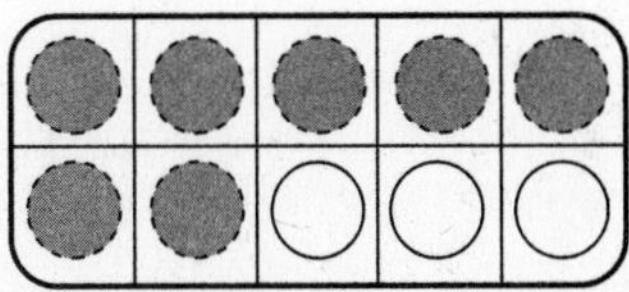

7 + 3 = 10

2. 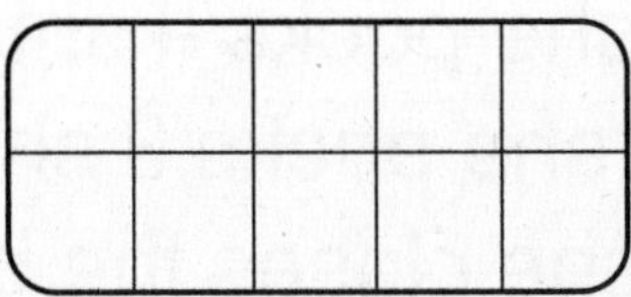

____ + ____ = ____

3. 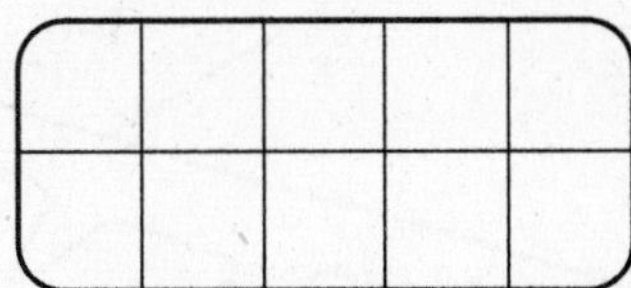

____ + ____ = ____

4. 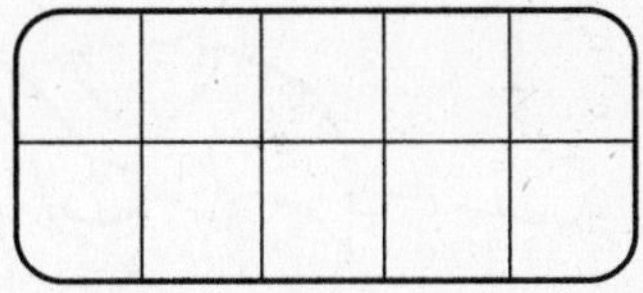

____ + ____ = ____

5. 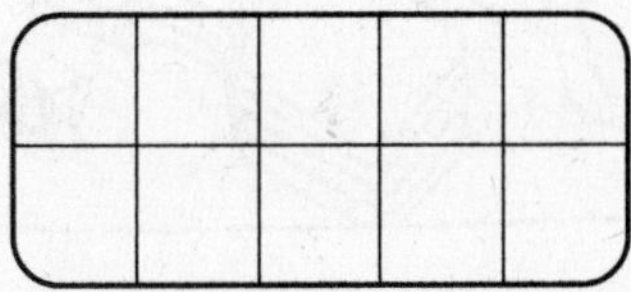

____ + ____ = ____

6. 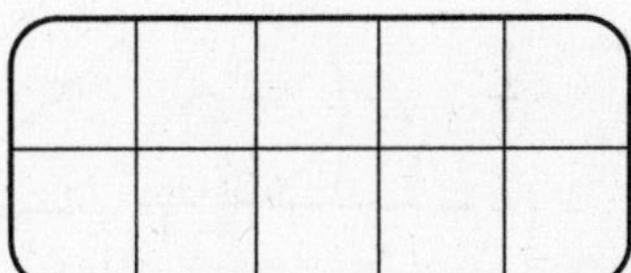

____ + ____ = ____

7. 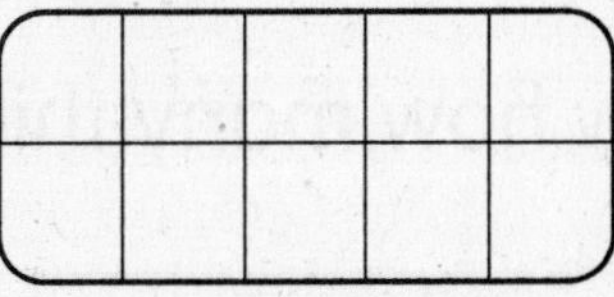

____ + ____ = ____

8. 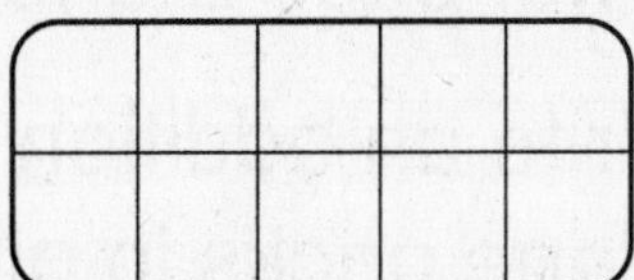

____ + ____ = ____

Name ______________________________

Problem Solving Skill: Reading for Math

Sequence of Events

Read the story.

Diane is packing her bag.
First she packs 4 shirts.
Then she packs 3 shorts.
Last she closes the bag.

Solve.

1. What does Diane do first? ______________________

2. What does Diane do last? ______________________

3. Write an addition sentence to show how many things Diane packs in all. ______________________

4. What if Diane packs 5 shirts and 4 shorts? Write an addition sentence to show how many things she would pack in all. ______________________

Use with Grade 1, Chapter 5, Lesson 4, pages 77–78.

Name ___

Add Across and Down

Write the numbers. Add.

1. 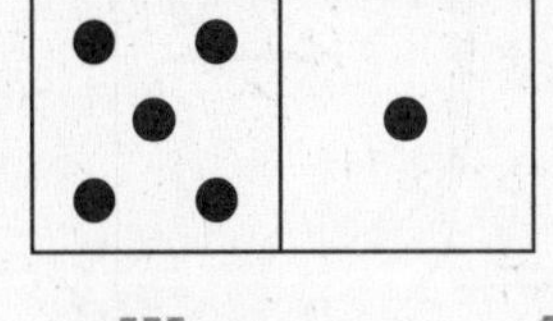

5 + 1 = 6

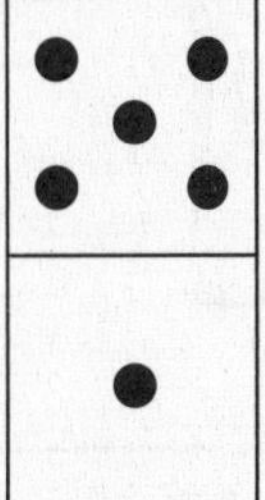

$$\begin{array}{r} 5 \\ +\ 1 \\ \hline 6 \end{array}$$

2. 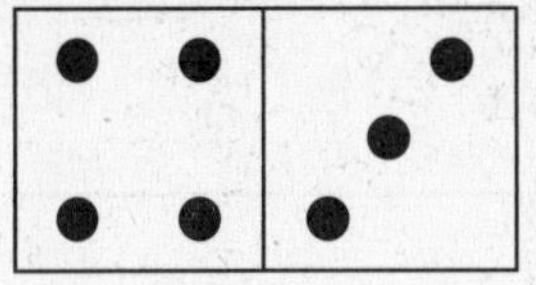

___ + ___ = ___

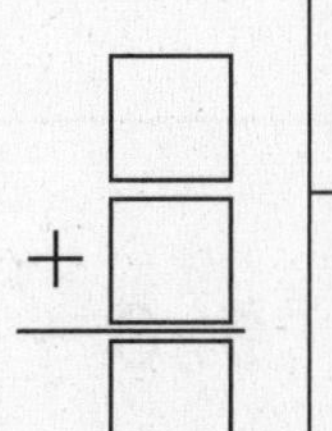

3. 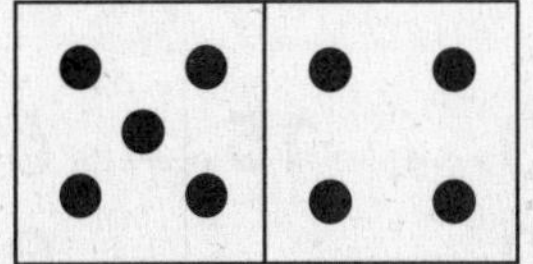

___ + ___ = ___

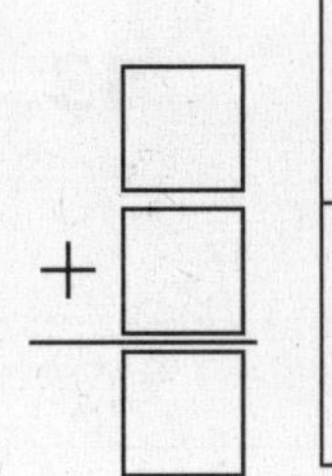

Name ____________________

Add 0

P 6-2 PRACTICE

Add.

1.

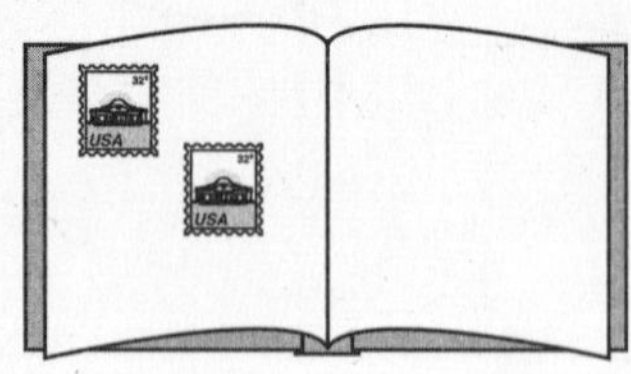

2 + 0 = 2

2.

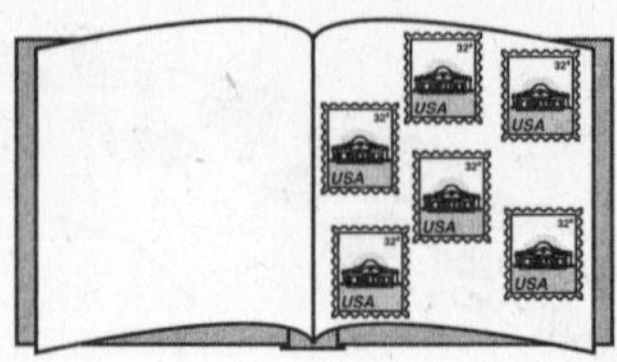

0 + 6 = ____

3.

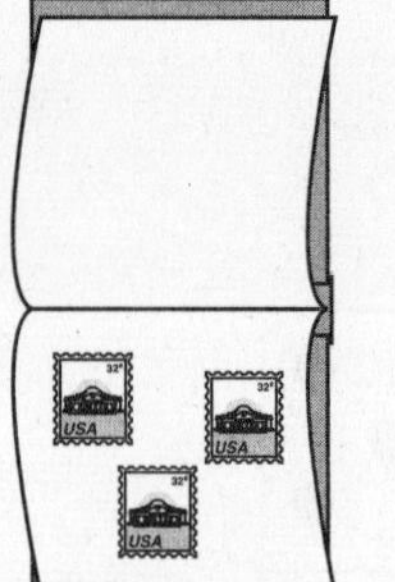

$$\begin{array}{r} 0 \\ +\ 3 \\ \hline \end{array}$$

4.

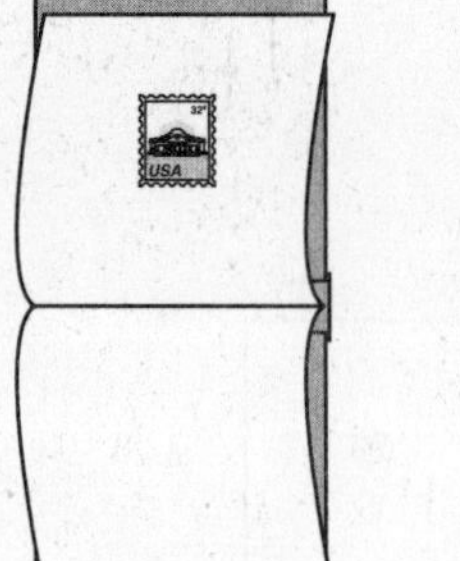

$$\begin{array}{r} 1 \\ +\ 0 \\ \hline \end{array}$$

5. 4 + 0 = ____ 6. 8 + 0 = ____ 7. 0 + 7 = ____

8. 1 + 0 = ____ 9. 0 + 2 = ____ 10. 5 + 0 = ____

11. $\begin{array}{r} 2 \\ +\ 0 \\ \hline \end{array}$ 12. $\begin{array}{r} 0 \\ +\ 4 \\ \hline \end{array}$ 13. $\begin{array}{r} 0 \\ +\ 8 \\ \hline \end{array}$ 14. $\begin{array}{r} 0 \\ +\ 0 \\ \hline \end{array}$ 15. $\begin{array}{r} 6 \\ +\ 0 \\ \hline \end{array}$

Name ________________________________

Add in Any Order

Write a fact for each picture.

You can use .

The addends are the same.
So the sums are the same.

1. 3 + 5 = 8

   ____ + ____ = ____

2. ____ + ____ = ____

   ____ + ____ = ____

3. ____ + ____ = ____

   ____ + ____ = ____

Add.

4. 6 + 3 = ____ 3 + 6 = ____

5. 1 + 5 = ____ 5 + 1 = ____

Name ____

Addition Practice

Add. Then color.

Sums of 8: red   Sums of 9: blue   Sums of 10: yellow

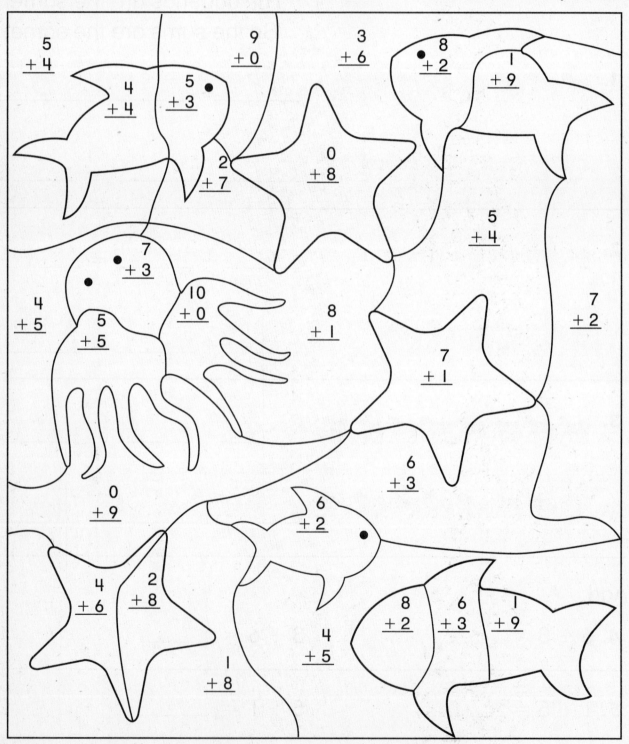

$5 + 4$

$9 + 0$

$3 + 6$

$8 + 2$

$1 + 9$

$4 + 4$

$5 + 3$

$2 + 7$

$0 + 8$

$5 + 4$

$7 + 3$

$10 + 0$

$4 + 5$

$5 + 5$

$8 + 1$

$7 + 2$

$7 + 1$

$6 + 3$

$0 + 9$

$6 + 2$

$4 + 6$

$2 + 8$

$8 + 2$

$6 + 3$

$1 + 9$

$4 + 5$

$1 + 8$

Name ______________________________

Problem Solving: Strategy

Draw a Picture

Draw a picture to solve.

1. Kayla read 4 .

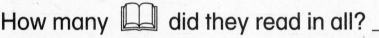

Miguel read 3 .

How many did they read in all? 7 

---

2. Ann sees 4 .

Then she sees 2 more  .

How many does she see? _____ 

---

3. Jane counted 3 .

Josh counted 5 .

How many did they count in all? _____

4. 6 are in the nest.

1 joins them.

How many  are in the nest? _____ 

Name ____________________

Subtract From 4, 5, and 6

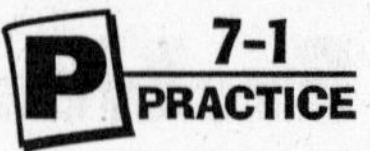

Use . Snap off some. Write the numbers.

	Subtract from 4, 5, and 6				
		minus		equals	difference
1.	4	−	____	=	____
2.	4	−	____	=	____
3.	4	−	____	=	____
4.	5	−	____	=	____
5.	5	−	____	=	____
6.	5	−	____	=	____
7.	6	−	____	=	____
8.	6	−	____	=	____
9.	6	−	____	=	____

Name ____________________

Subtract From 7, 8, and 9

Use . Snap off some. Write the numbers.

	Subtract from 7, 8, and 9				
		minus		equals	difference
1.	7	−	____	=	____
2.	7	−	____	=	____
3.	7	−	____	=	____
4.	8	−	____	=	____
5.	8	−	____	=	____
6.	8	−	____	=	____
7.	9	−	____	=	____
8.	9	−	____	=	____
9.	9	−	____	=	____

Name ____

Subtract From 10

P 7-3 PRACTICE

Complete the subtraction sentences.

1. 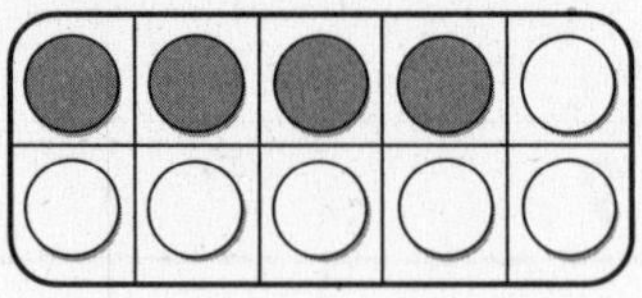 $10 - 6 =$ 4

   $10 - 4 =$ ______

2. 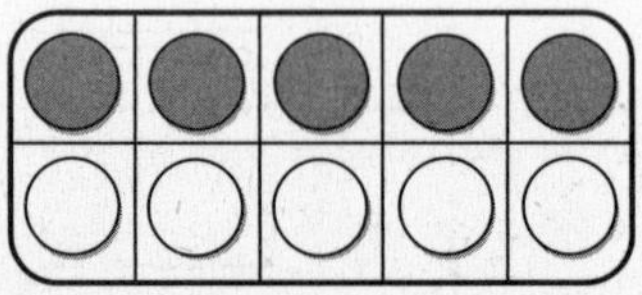 $10 - 5 =$ ______

3. 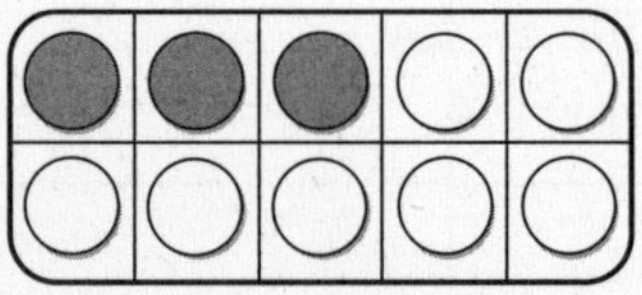 $10 - 7 =$ ______

   $10 - 3 =$ ______

4. 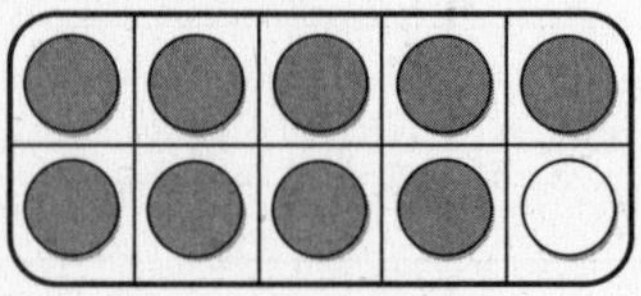 $10 - 1 =$ ______

   $10 - 9 =$ ______

5. 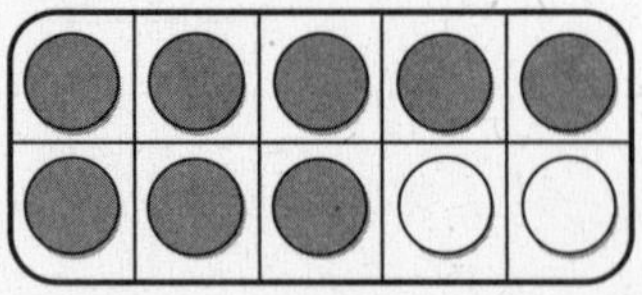 $10 - 2 =$ ______

   $10 - 8 =$ ______

Use with Grade 1, Chapter 7, Lesson 3, pages 107–108.

Name ___

Missing Numbers • Algebra

Subtract to find the missing numbers.
Draw the dots.

1. 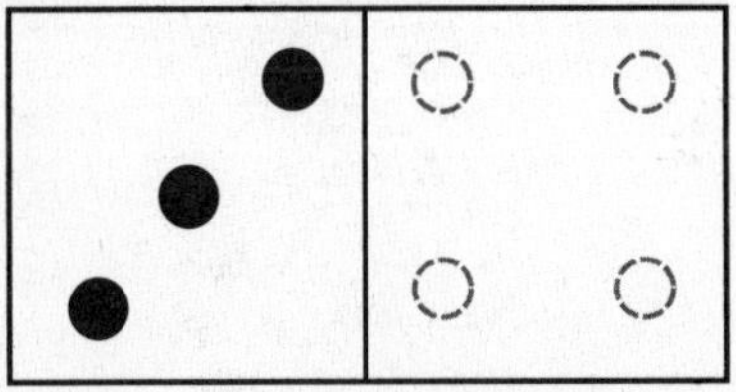

7 − 3 = 4

2. 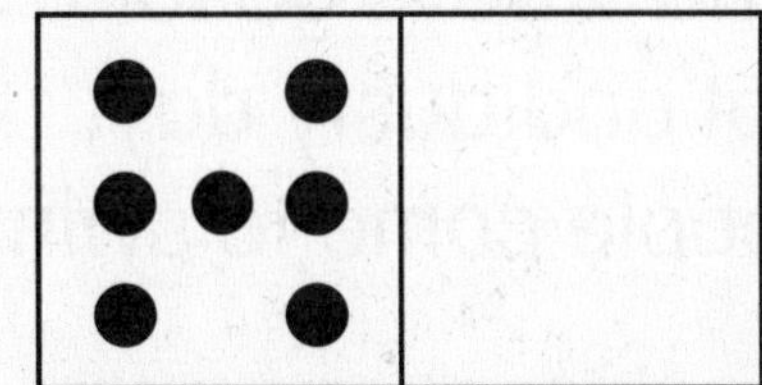

10 − 7 = ___

3. 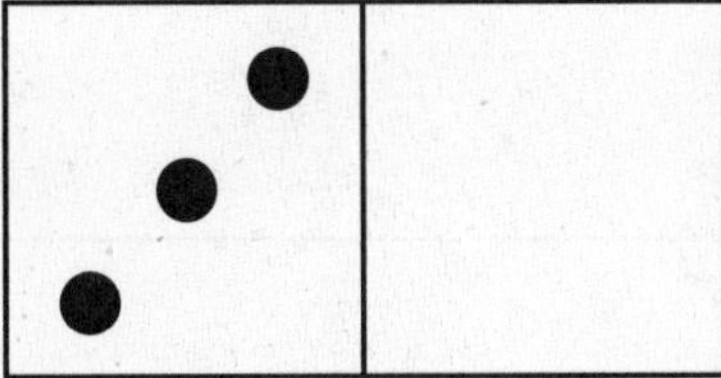

6 − 3 = ___

4. 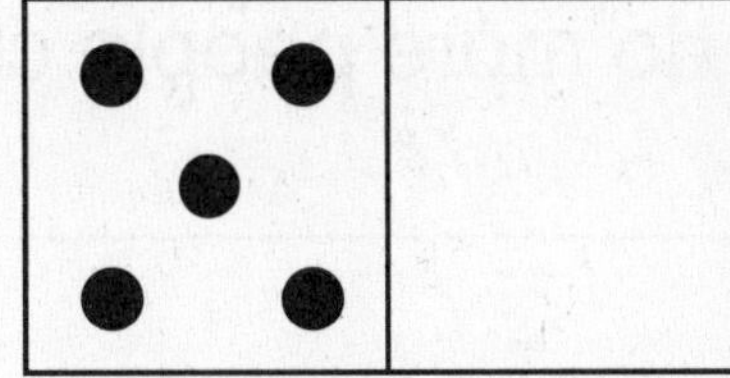

7 − 5 = ___

5. 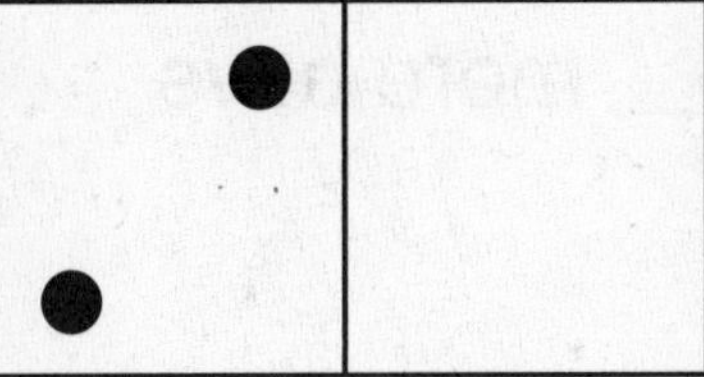

5 − 2 = ___

6. 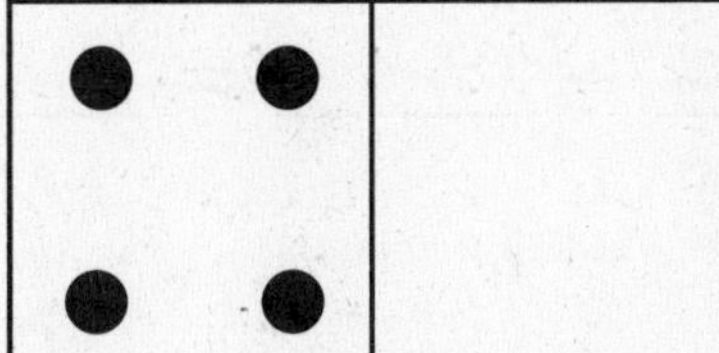

10 − 4 = ___

7. 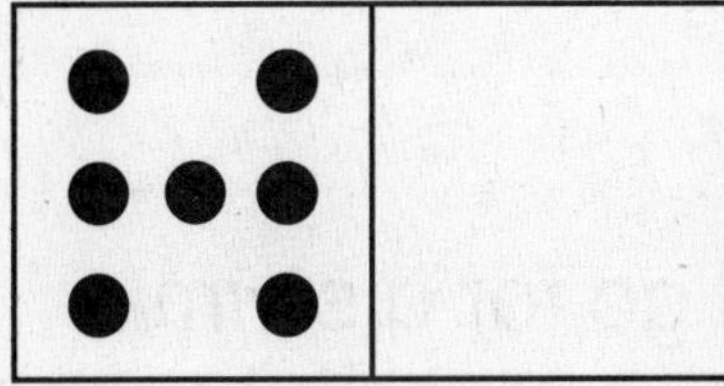

8 − 7 = ___

8. 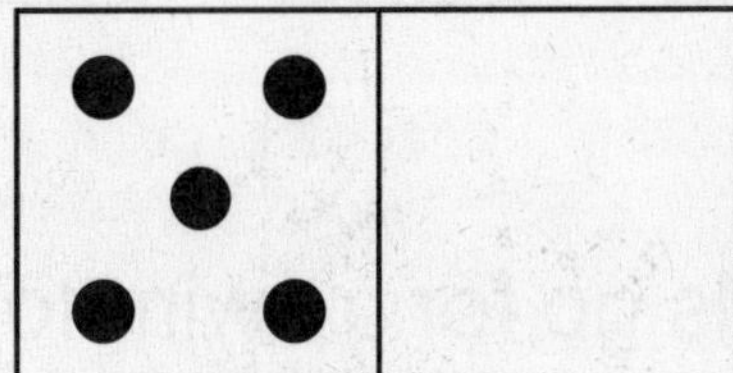

9 − 5 = ___

Name ______________________________

Problem Solving Skill: Reading for Math

Cause and Effect

The farm has a big pond.
7 girls and 9 boys go for a swim.
It is a hot and dusty day.
More people come to swim.

Solve.

1. Why do more people come to swim?

____

2. How many more boys swim than girls? Write a subtraction sentence.

_____ − _____ = _____ _____ more boys

3. 9 boys go for a swim today. 4 boys went for a swim yesterday. Subtract to show the difference.

_____ − _____ = _____

4. 7 girls go for a swim today. 5 girls will go for a swim tomorrow. Subtract to show the difference.

_____ − _____ = _____

Name ____

Subtract Across and Down

Cross out to subtract.

1. 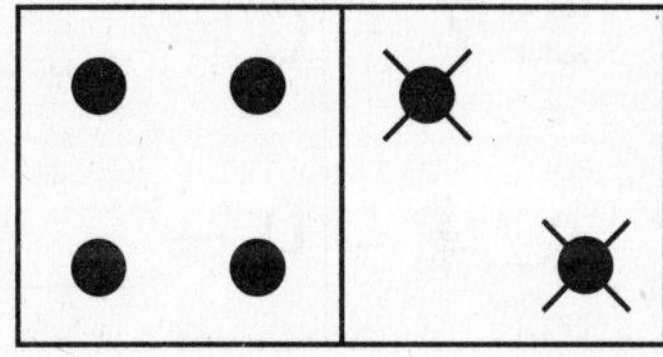

6 − 2 = 4

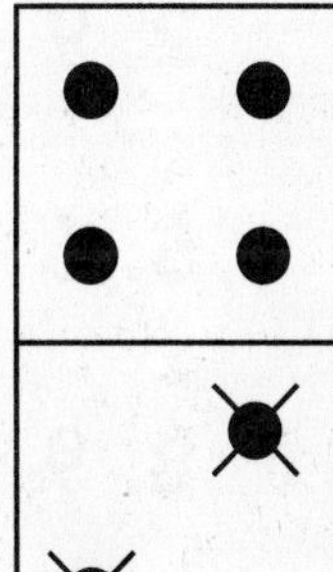

$$\begin{array}{r} 6 \\ -\ 2 \\ \hline 4 \end{array}$$

2. 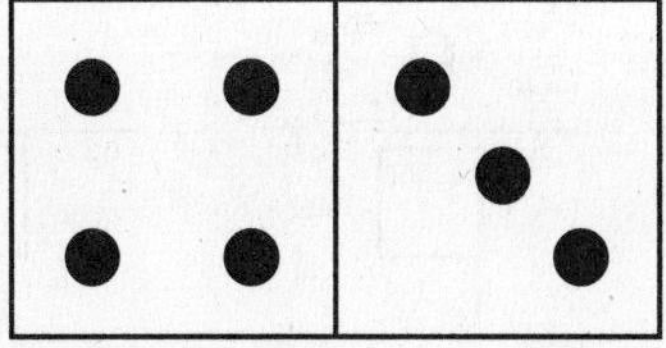

7 − 4 = ______

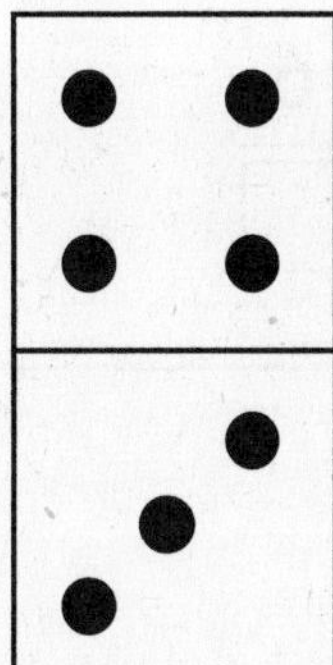

$$\begin{array}{r} 7 \\ -\ 4 \\ \hline \end{array}$$

3. 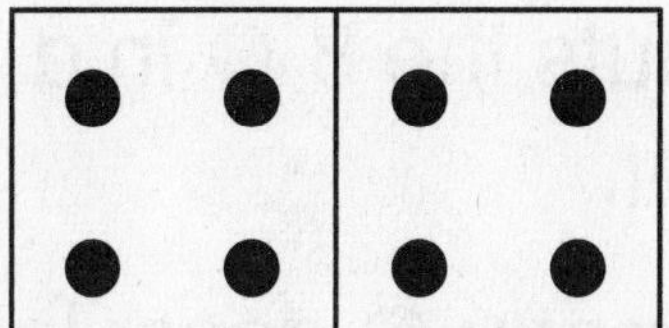

8 − 4 = ______

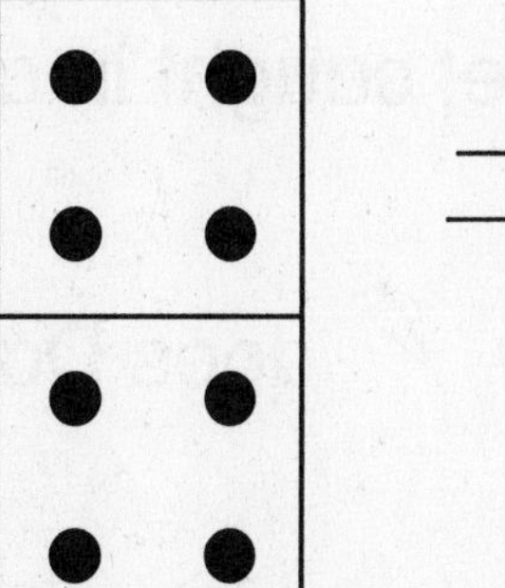

$$\begin{array}{r} 8 \\ -\ 4 \\ \hline \end{array}$$

4. 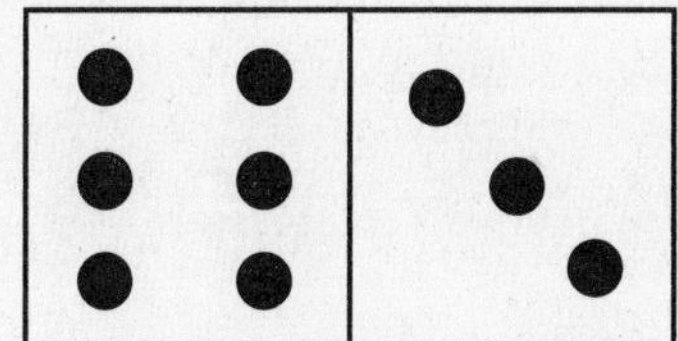

9 − 3 = ______

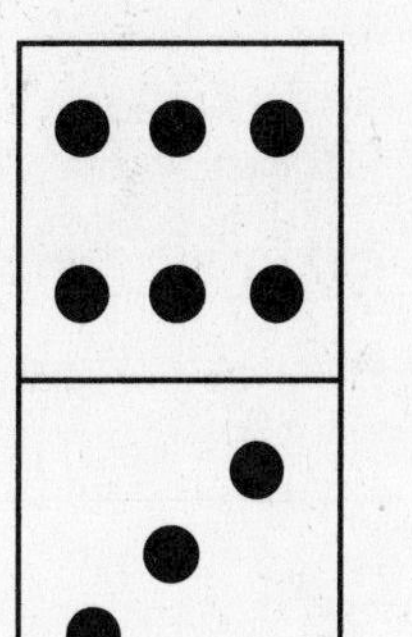

$$\begin{array}{r} 9 \\ -\ 3 \\ \hline \end{array}$$

Name ____________________

Subtract 0 and Subtract All

P 8-2 PRACTICE

Subtract.

1. 8 − 0 = 8

2. 9 − 9 = ____

3. 7 − 0 = ____

4. 5 − 5 = ____

5. 6 − 0 = ____

6. 4 − 4 = ____

7. 3 − 3 = ____

8. 9 − 0 = ____

9. 7 − 7 = ____

---

10. $\begin{array}{r} 4 \\ -\ 0 \\ \hline \square \end{array}$ **11.** $\begin{array}{r} 6 \\ -\ 6 \\ \hline \square \end{array}$ **12.** $\begin{array}{r} 8 \\ -\ 8 \\ \hline \square \end{array}$ **13.** $\begin{array}{r} 3 \\ -\ 0 \\ \hline \square \end{array}$ **14.** $\begin{array}{r} 5 \\ -\ 0 \\ \hline \square \end{array}$

Problem Solving

15. Dan has 4 .

All 4 get caught in a tree.

How many does Dan have left?

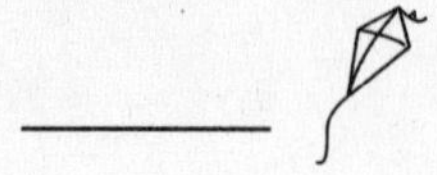

16. Jerri has 8 .

She puts the 8 in a basket.

How many does Jerri have?

Name ____________________

Use Addition to Check Subtraction • Algebra

Subtract. Check by adding.

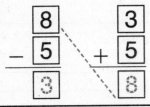

1.
$$\begin{array}{r} \boxed{8} \\ -\ \boxed{5} \\ \hline \boxed{3} \end{array} \qquad \begin{array}{r} \boxed{3} \\ +\ \boxed{5} \\ \hline \boxed{8} \end{array}$$

2.
$$\begin{array}{r} \boxed{7} \\ -\ \boxed{2} \\ \hline \boxed{} \end{array} \qquad \begin{array}{r} \boxed{5} \\ +\ \boxed{2} \\ \hline \boxed{} \end{array}$$

3.
$$\begin{array}{r} \boxed{6} \\ -\ \boxed{4} \\ \hline \boxed{} \end{array} \qquad \begin{array}{r} \boxed{2} \\ +\ \boxed{4} \\ \hline \boxed{} \end{array}$$

4.
$$\begin{array}{r} \boxed{5} \\ -\ \boxed{3} \\ \hline \boxed{} \end{array} \qquad \begin{array}{r} \boxed{2} \\ +\ \boxed{3} \\ \hline \boxed{} \end{array}$$

5.
$$\begin{array}{r} 6 \\ -\ 5 \\ \hline \boxed{} \end{array} \qquad \begin{array}{r} 1 \\ +\ 5 \\ \hline \boxed{} \end{array}$$

6.
$$\begin{array}{r} 8 \\ -\ 2 \\ \hline \boxed{} \end{array} \qquad \begin{array}{r} 6 \\ +\ 2 \\ \hline \boxed{} \end{array}$$

Problem Solving

Solve. Check by adding.

7. Tom has 7  .

He loses 1  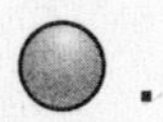 .

How many ● does Tom have now?

______ ●

____

____

8. Scott has 6  .

He gives 1 ● away.

How many ● does Scott have now?

____

____

Name ____________________

Fact Families

Use cubes if you like. Add and subtract. Write the numbers in the fact family.

1. 3 + 6 = 9   9 − 6 = 3   3, 6, 9

6 + 3 = 9   9 − 3 = 6

2. 2 + 5 = ____   7 − 2 = ____   ____, ____, ____

5 + 2 = ____   7 − 5 = ____

3. 3 + 5 = ____   8 − 5 = ____   ____, ____, ____

5 + 3 = ____   8 − 3 = ____

4. 8 + 2 = ____   10 − 2 = ____   ____, ____, ____

2 + 8 = ____   10 − 8 = ____

Problem Solving

Solve.

5. There are 10 balls.

The children take 5 balls.

How many balls are left?

____ balls left

6. There are 5 red balls.

There are 5 blue balls.

How many balls are there in all?

____ balls in all

Use with Grade 1, Chapter 8, Lesson 4, pages 125–126.

Name ____________________

Problem Solving: Strategy

Act It Out • Algebra

Subtract. Use ◯ to solve.

1. There are 6 🐶. 2 🐶 are playing. How many 🐶 are not playing?

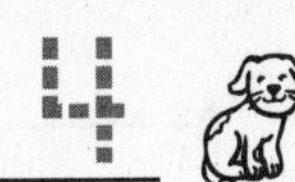

2. There were 4 🐱. 2 🐱 hid under the bed. How many 🐱 are left?

3. There were 7 👧. 3 👧 got out of the pool. How many 👧 are still in the pool?

4. There are 8 🦆. 6 🦆 swim away. How many 🦆 did not swim away?

Name ____________________

Count On 1 or 2

P 9-1 PRACTICE

Use the number line to add.
Count on 1 or 2.

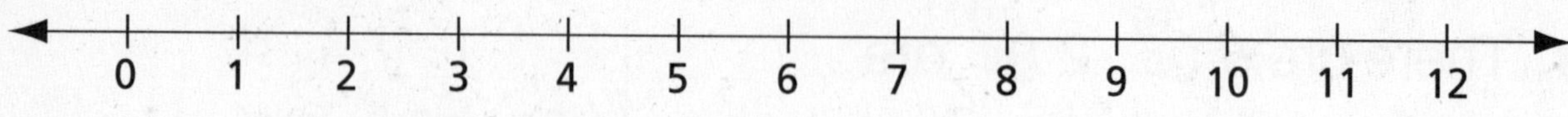

1. 9 + 2 = 11 2. 8 + 1 = ____ 3. 6 + 1 = ____

4. 7 + 2 = ____ 5. 5 + 1 = ____ 6. 4 + 2 = ____

7. 7 + 1 = ____ 8. 5 + 2 = ____ 9. 6 + 2 = ____

10. 8 + 2 = ____ 11. 9 + 1 = ____ 12. 3 + 2 = ____ 13. 2 + 2 = ____ 14. 4 + 1 = ____ 15. 3 + 1 = ____

Problem Solving

Solve.

Show Your Work

16. 3 flowers are in a vase.

Draw 1 more.

How many flowers are in the vase now?

____ flowers

Use with Grade 1, Chapter 9, Lesson 1, pages 143–144.

Name ____________________

Count On 1, 2, or 3

Circle the greater number. Count on 1, 2, or 3.

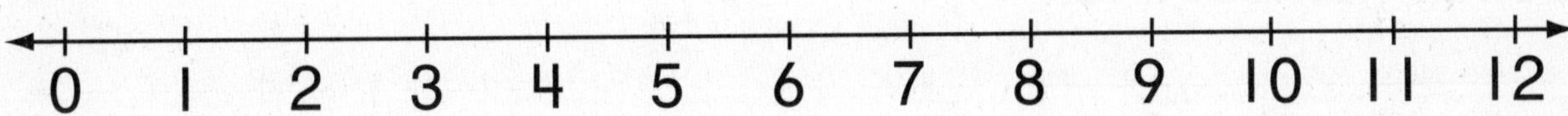

1. (5) + 3 = 8

2. 6 + 1 = ____

3. 2 + 2 = ____

4. 9 + 2 = ____

5. 8 + 2 = ____

6. 7 + 3 = ____

7. 6 + 2 = ____

8. 9 + 3 = ____

9. 7 + 2 = ____

10. $\begin{array}{r} 8 \\ +1 \\ \hline \end{array}$

11. $\begin{array}{r} 6 \\ +3 \\ \hline \end{array}$

12. $\begin{array}{r} 7 \\ +2 \\ \hline \end{array}$

13. $\begin{array}{r} 9 \\ +1 \\ \hline \end{array}$

14. $\begin{array}{r} 8 \\ +2 \\ \hline \end{array}$

15. $\begin{array}{r} 8 \\ +3 \\ \hline \end{array}$

Circle the greater number. Count on 1, 2, or 3.

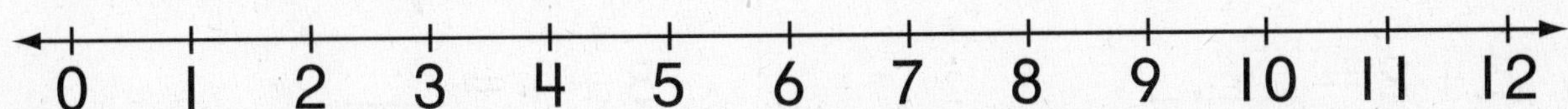

16. 8 + 3 = ____ N

17. 6 + 3 = ____ W

18. 5 + 2 = ____ H

19. 9 + 3 = ____ D

20. 6 + 2 = ____ E

21. 5 + 1 = ____ T

22. 8 + 2 = ____ I

Answer the riddle. Write the letters that match the sums.

What can you hear but not see?

____	____	____		____	____	____	____
6	7	8		9	10	11	12

Name ____

Estimate Sums

9-3 PRACTICE

Estimate the sum. Add to check.
Use the number line.

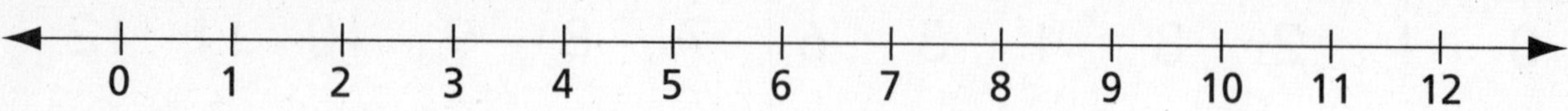

1. 6 + 2

more than 10

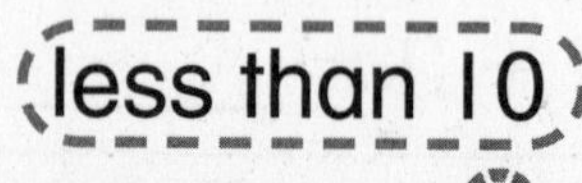

less than 10

6 + 2 = 8

2. 7 + 3

more than 8

less than 8

7 + 3 = ______

3. 5 + 2

more than 9

less than 9

5 + 2 = ______

4. 8 + 1

more than 10

less than 10

8 + 1 = ______

Make it Right

5. Jason counted on like this.

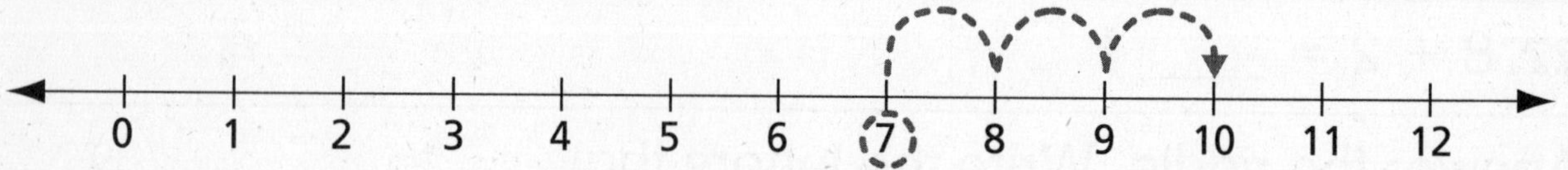

7 + 2 = 10

Why is Jason wrong? Make it right.

____

Name ________________________________

Ways to Make 11 and 12

9-4 PRACTICE

Draw ● and ○ on ⊞.

Show ways to make 11.

Write each addition sentence.

1. 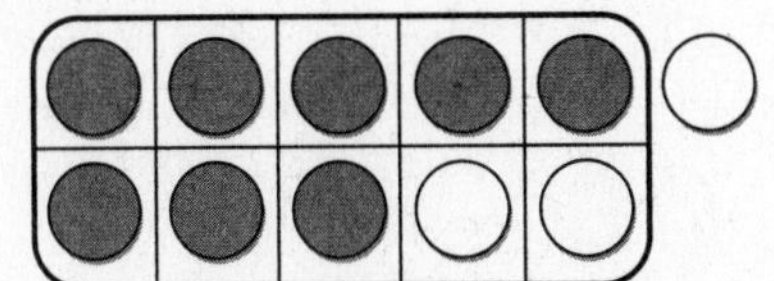

8 + 3 = 11

2. 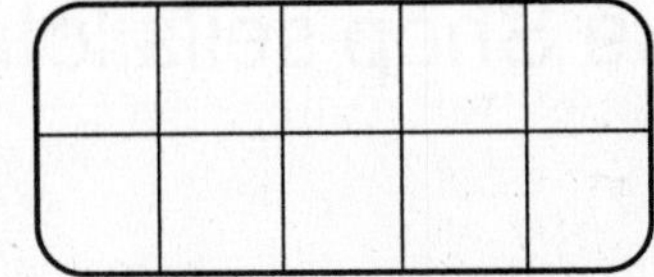

____ + ____ = ____

3. 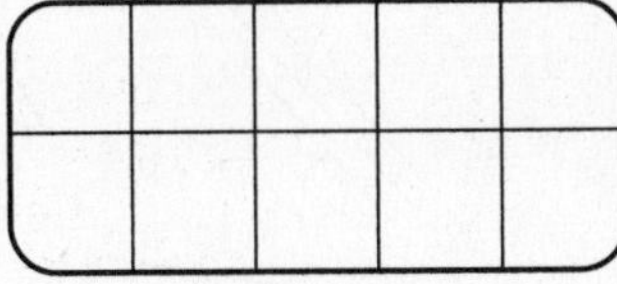

____ + ____ = ____

4. 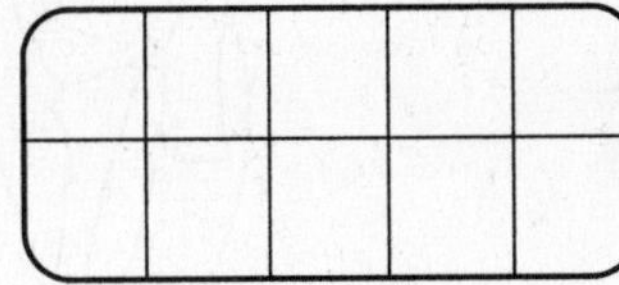

____ + ____ = ____

Show ways to make 12.

Write each addition sentence.

5. 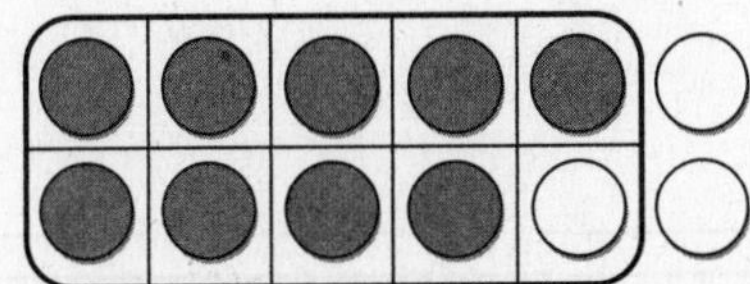

9 + 3 = 12

6. 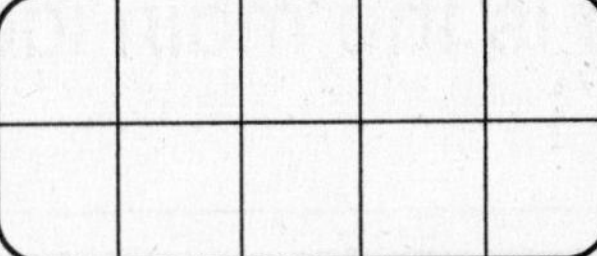

____ + ____ = ____

7. 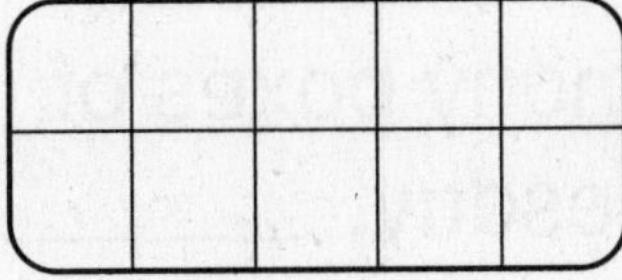

____ + ____ = ____

8. 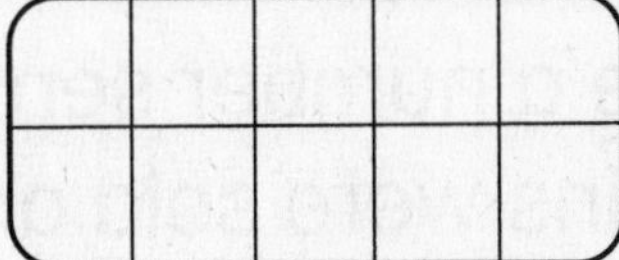

____ + ____ = ____

Name ____________________

Problem Solving Skill: Reading for Math

Find the Main Idea

Muffins taste good!
The Bake Shop sells 5 boxes of muffins on Monday.
It sells 3 boxes on Tuesday.
The Bake Shop sells lots of muffins.

1. What is the main idea of the story?

____

2. On what day did the Bake Shop sell the most muffins?

____

3. Write a number sentence. Find how many boxes of muffins were sold on Monday and Tuesday. ____________

4. How many boxes were sold in all? ______ boxes

Name ________________________________

Count Back 1 or 2

Count back to subtract. Use the number line.

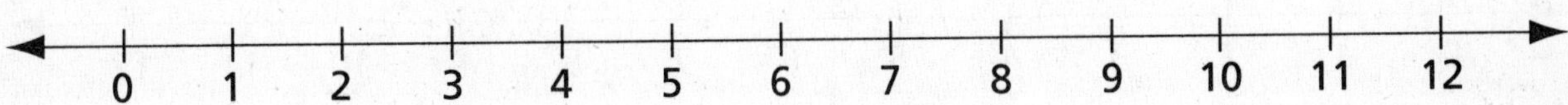

1. 7 − 1 = 6

2. 6 − 2 = ______

3. 6 − 1 = ______

4. 9 − 2 = ______

5. 10 − 1 = ______

6. 8 − 2 = ______

7. 7 − 2 = ______

8. 9 − 1 = ______

9. 3 − 2 = ______

10. $\begin{array}{r} 2 \\ -\ 1 \\ \hline \end{array}$

11. $\begin{array}{r} 11 \\ -\ 2 \\ \hline \end{array}$

12. $\begin{array}{r} 1 \\ -\ 1 \\ \hline \end{array}$

13. $\begin{array}{r} 4 \\ -\ 2 \\ \hline \end{array}$

14. $\begin{array}{r} 5 \\ -\ 2 \\ \hline \end{array}$

15. $\begin{array}{r} 10 \\ -\ 2 \\ \hline \end{array}$

Problem Solving

Solve.

Draw a picture to solve.

16. Ted has 8 🍎.
He eats 1 🍎.
How many 🍎 are left?
______ 🍎 are left.

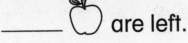

Name ______________________________

Count Back 1, 2, or 3

10-2 PRACTICE

Count back to subtract.

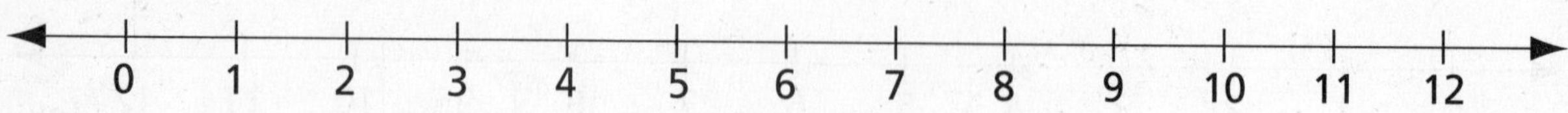

1. 11 − 3 = 8

2. 9 − 2 = ______

3. 3 − 3 = ______

4. 8 − 1 = ______

5. 4 − 2 = ______

6. 12 − 3 = ______

7. 5 − 3 = ______

8. 7 − 2 = ______

9. 7 − 3 = ______

10. 4 − 3 = ______

11. 11 − 2 = ______

12. 6 − 3 = ______

13. 6 − 1 = ______

14. 5 − 2 = ______

15. 10 − 3 = ______

Problem Solving

Solve.

Circle the correct number sentence.

16. Darla has 5 .

She mails 3 .

How many are left?

______  are left.

8 − 5 = 3

3 − 5 = 2

5 − 3 = 2

3 + 2 = 5

Name ______________________________

Estimate Differences

Estimate the difference. Subtract to check.
Use the number line.

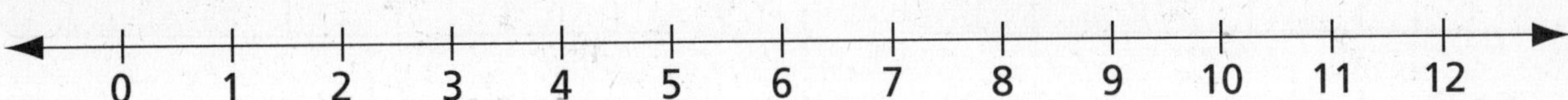

1. 9 − 2

more than 5

less than 5

9 − 2 = 7

2. 6 − 1

more than 4

less than 4

6 − 1 = ______

3. 12 − 3

more than 10

less than 10

12 − 3 = ______

4. 10 − 3

more than 6

less than 6

10 − 3 = ______

Make It Right

5. Myra subtracts 5 − 2 like this.

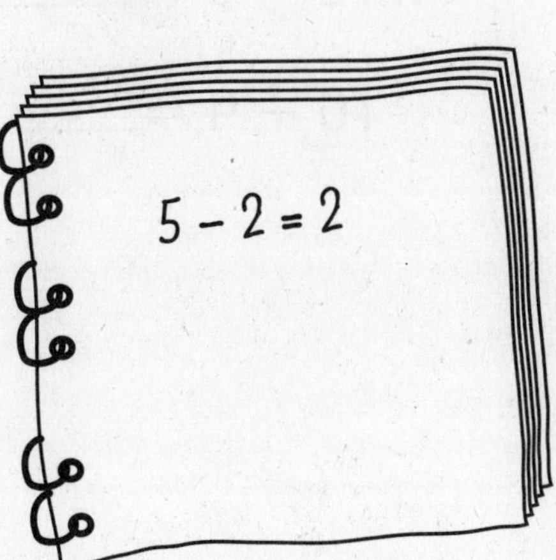

Why is Myra wrong?
Make it right.

____

____

Name ______________________________

Related Subtraction Facts

Subtract.

1. 11 − 6 = 5

11 − 5 = 6

2. 12 − 8 = ____

12 − 4 = ____

3. 11 − 7 = ____

11 − 4 = ____

4. 10 − 3 = ____

10 − 7 = ____

5. 12 − 3 = ____

12 − 9 = ____

6. 9 − 1 = ____

9 − 8 = ____

7. 7 − 2 = ____

7 − 5 = ____

8. 5 − 3 = ____

5 − 2 = ____

9. 8 − 2 = ____

8 − 6 = ____

10. 4 − 3 = ____

4 − 1 = ____

11. 8 − 3 = ____

8 − 5 = ____

12. 6 − 1 = ____

6 − 5 = ____

13. 7 − 3 = ____

7 − 4 = ____

14. 10 − 6 = ____

10 − 4 = ____

Name ____________________

Practice the Facts

Riddle: What is the first thing you see in ANIMALS?

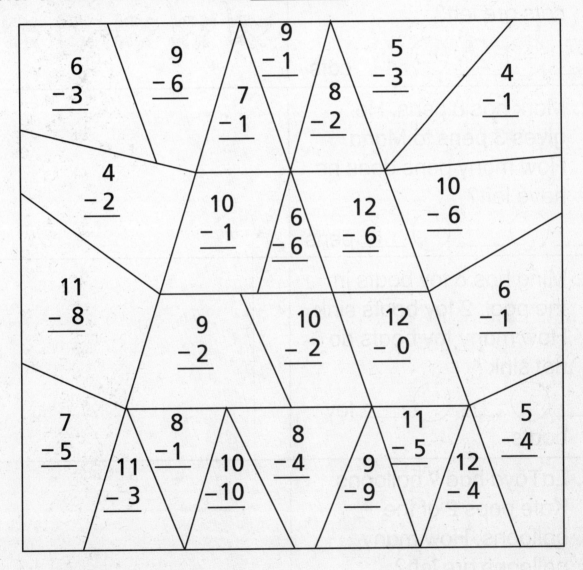

Add or subtract.
Color answers of 6 or greater blue.
Color answers of 5 or less yellow.

$6 - 3$ | $9 - 6$ | $9 - 1$ | $7 - 1$ | $8 - 2$ | $5 - 3$ | $4 - 1$

$4 - 2$ | $10 - 1$ | $6 - 6$ | $12 - 6$ | $10 - 6$

$11 - 8$ | $9 - 2$ | $10 - 2$ | $12 - 0$ | $6 - 1$

$7 - 5$ | $8 - 1$ | $11 - 3$ | $10 - 10$ | $8 - 4$ | $9 - 9$ | $11 - 5$ | $12 - 4$ | $5 - 4$

Complete the answer.

You see ________.

Name ____________________

Problem Solving: Strategy

Write a Number Sentence

Write a number sentence to solve. Draw or write to explain.

1. 5 cats are on the mat. 2 cats run away. How many cats are left?

cats

2. Mark has 8 pens. He gives 3 pens to Maria. How many pens does he have left?

___ ◯ ___ ◯ ___ pens

3. Ming has 6 toy boats in the pool. 2 toy boats sink. How many toy boats do not sink?

___ ◯ ___ ◯ ___ toy boats

4. LaToya had 9 balloons. Kate pops 2 of the balloons. How many balloons are left?

___ ◯ ___ ◯ ___ balloons

Use with Grade 1, Chapter 10, Lesson 6, pages 169–170.

Name ______________________________

Relate Addition to Subtraction

Add. Write the related subtraction facts.

1. 7 + 3 = 10
10 ⊖ 3 = 7
10 ⊖ 7 = 3

2. 2 + 6 = ____
____ ◯ ____ = ____
____ ◯ ____ = ____

3. 9 + 2 = ____
____ ◯ ____ = ____
____ ◯ ____ = ____

4. 10 + 2 = ____
____ ◯ ____ = ____
____ ◯ ____ = ____

5. 6 + 5 = ____
____ ◯ ____ = ____
____ ◯ ____ = ____

6. 5 + 4 = ____
____ ◯ ____ = ____
____ ◯ ____ = ____

7. 5 + 3 = ____
____ ◯ ____ = ____
____ ◯ ____ = ____

8. 7 + 4 = ____
____ ◯ ____ = ____
____ ◯ ____ = ____

9. 8 + 4 = ____
____ ◯ ____ = ____
____ ◯ ____ = ____

10. 6 + 3 = ____
____ ◯ ____ = ____
____ ◯ ____ = ____

Name ____________________

Doubles Facts

P 11-2 PRACTICE

Use doubles to add and subtract.

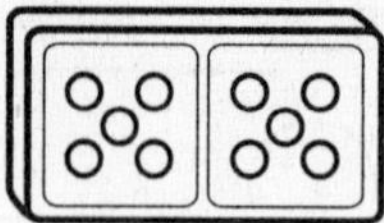

1. 5 + 5 = 10

10 − 5 = 5

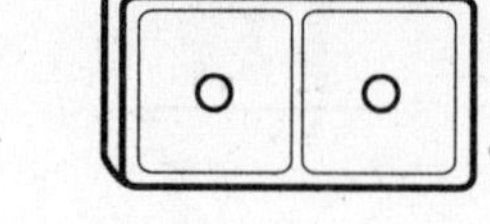

2. 1 + 1 = ________

2 − 1 = ________

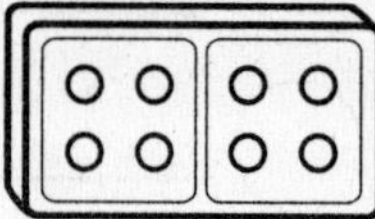

3. 4 + 4 = ________

8 − 4 = ________

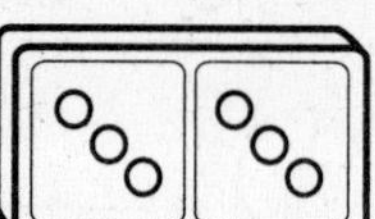

4. 3 + 3 = ________

6 − 3 = ________

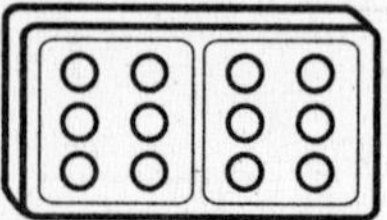

5. 6 + 6 = ________

12 − 6 = ________

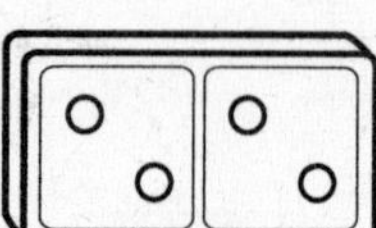

6. 2 + 2 = ________

4 − 2 = ________

Name ______________________________

Missing Addend • Algebra

P 11-3 PRACTICE

Find the missing addend. Use ◯.

1. 4 + [8] = 12

2. 1 + [ ] = 8

3. [ ] + 7 = 10

4. 5 + [ ] = 7

5. [ ] + 4 = 9

6. 2 + [ ] = 11

7. [ ] + 8 = 10

8. [ ] + 2 = 9

9. 3 + [ ] = 5

10. 3 + [ ] = 6

11. 9 + [ ] = 12

12. [ ] + 3 = 11

13. 5 + [ ] = 8

14. [ ] + 5 = 8

Name ______________________________

Fact Families to 12

11-4 PRACTICE

Add. Then subtract.
Complete each fact family.

1. 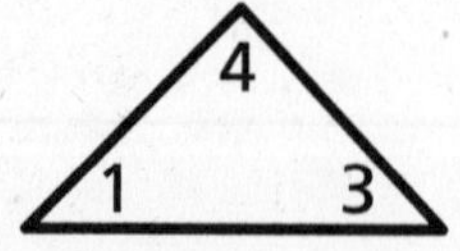

1 + 3 = 4 4 − 3 = ____
3 + 1 = ____ 4 − 1 = ____

---

2. 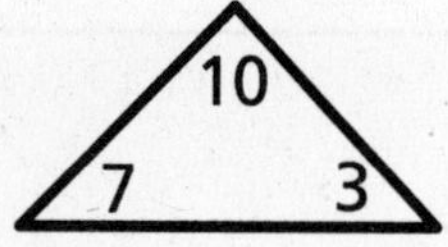

7 + 3 = ____ 10 − 3 = ____
3 + 7 = ____ 10 − 7 = ____

---

3. 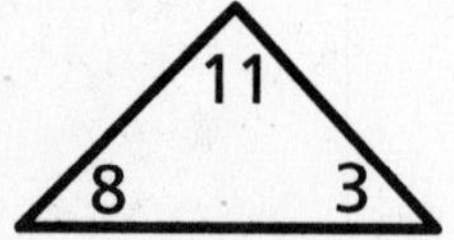

8 + 3 = ____ 11 − 3 = ____
3 + 8 = ____ 11 − 8 = ____

---

4. 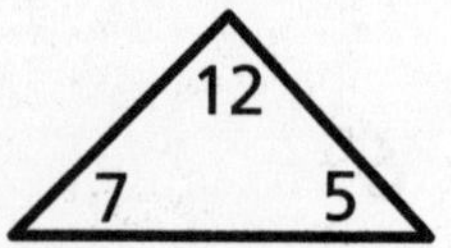

7 + 5 = ____ 12 − 5 = ____
5 + 7 = ____ 12 − 7 = ____

---

5. 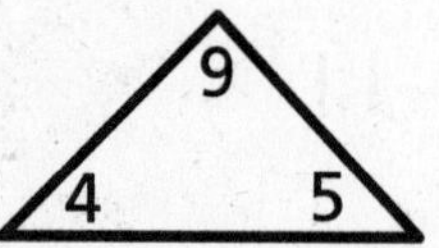

4 + 5 = ____ 9 − 5 = ____
5 + 4 = ____ 9 − 4 = ____

Name ________________________________

Ways to Name Numbers

Read each number.
Circle the ways to make that number.

1. 

9 + 2   7 + 4
10 − 1   6 + 5

2. 

4 + 2   7 − 2
10 − 5   9 − 4

3. 

7 + 3   12 − 3
4 + 6   1 + 9

4. 

10 − 2   9 − 2
2 + 5   0 + 7

5. 

4 + 4   3 + 6
9 − 1   1 + 7

6. 

2 + 9   3 + 8
4 + 7   11 − 1

7. 

2 + 3   8 − 4
10 − 6   3 + 1

8. 

7 + 4   5 + 7
3 + 9   9 + 3

9. 

11 − 5   4 + 2
6 − 0   12 − 7

10. 

12 − 3   1 + 8
9 − 1   9 − 0

Name ______________________________

Problem Solving Skill: Reading for Math

Find the Main Idea

There are lots of things to do at the beach.
3 children are playing ball.
1 child is feeding the birds.

1. What is the main idea of the story? ______________________

____

2. How many children are on the beach?

_____ children

Write the addition fact that shows the number of children.

_____ + _____ = _____

3. Write 2 related subtraction facts for the addition fact you just wrote.

_____ − _____ = _____

_____ − _____ = _____

Use with Grade 1, Chapter 11, Lesson 6, pages 187–188.

Name ______________________________

Read a Bar Graph

Use the bar graph. Answer the questions.

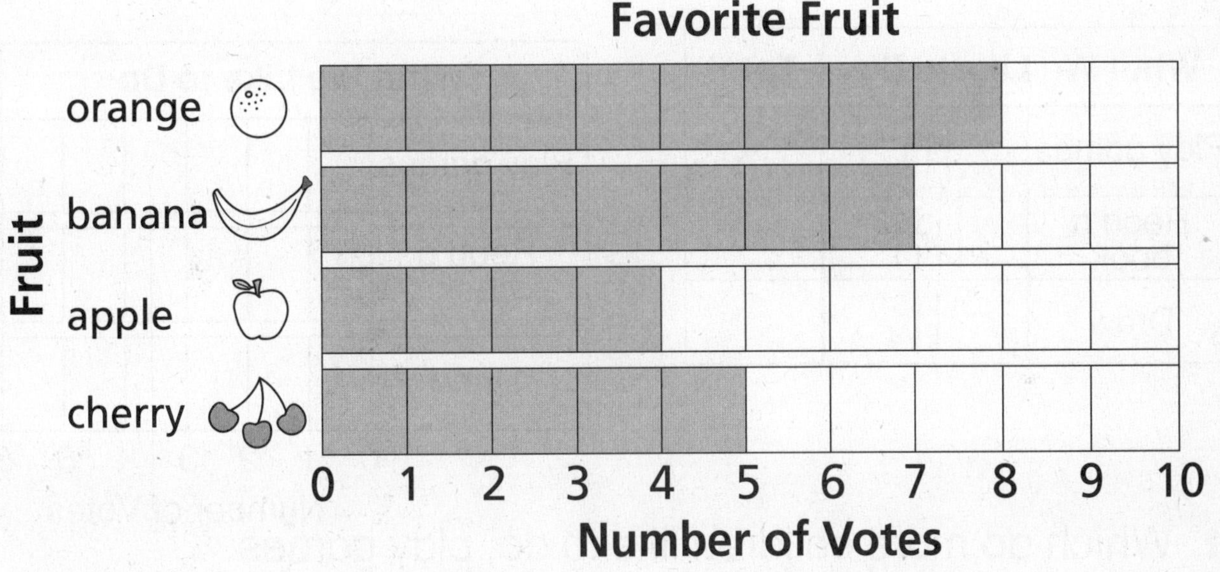

1. Which fruit has fewer votes, banana or cherry?

   cherry

2. Count the votes for apple and orange. How many votes in all? ______

3. Which fruit has fewer than 5 votes? ______________

4. How many more votes for orange than for banana?

   ______

5. Which fruit got the most votes? ______________

6. How many more votes for banana than for apple? ______

Name ____________________

Make a Bar Graph

Write each total. Make a bar graph.
Answer the questions.

<table>
<tr><th colspan="2">What We Like to Do</th><th>Total</th></tr>
<tr><td>Play games</td><td>|||</td><td>3</td></tr>
<tr><td>Read a book</td><td>||||/</td><td></td></tr>
<tr><td>Draw</td><td>||</td><td></td></tr>
</table>

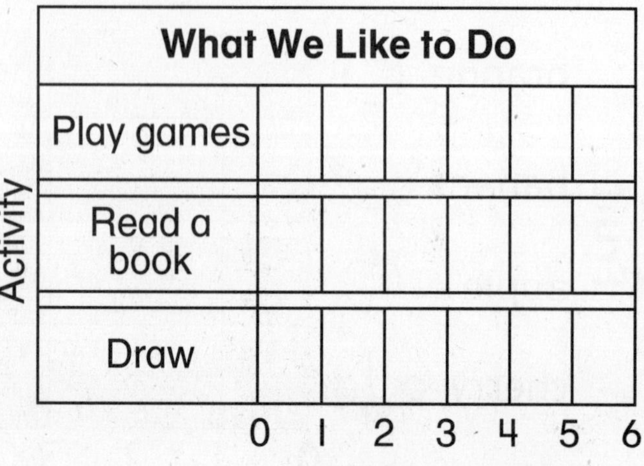

1. Which do more children like to do, play games or draw ? ____________________

2. Which activity got the most votes? ____________________

3. Which activity got the fewest votes? ____________________

4. Which 2 activities got 8 votes in all? ____________________

____

5. Which activity got fewer votes than play games? ____________________

Algebra Patterns

Write the number that comes next in the pattern.

6.

5	6	5	6	5	6	5	

7.

2	1	2	1	2	1	2	

Name ______________________________

Use a Bar Graph

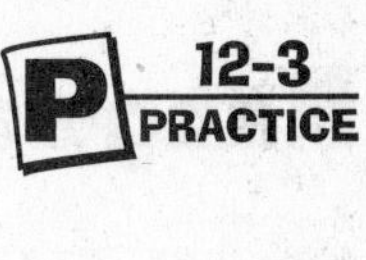

Use the bar graph to answer the questions.

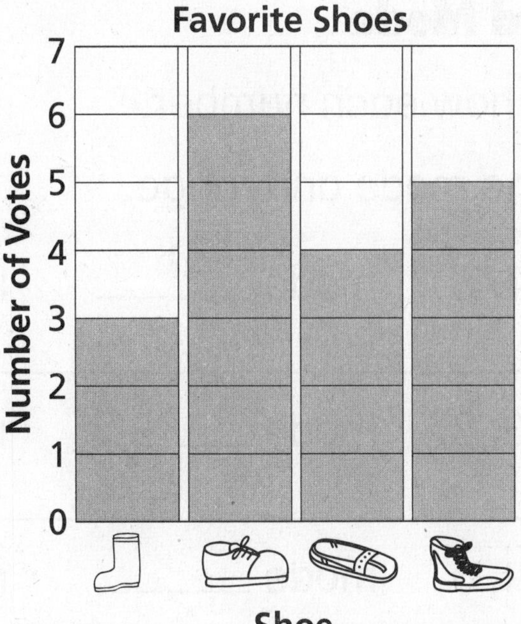

1. How many like [boot] the best? 3

2. How many votes did [slipper] and [sneaker] get in all? ______

3. How many more like [slipper] than [boot] ? ______

4. Which shoe got the most votes?
   Circle your answer.

5. Which type of shoe got 2 more votes than [boot] ? Circle your answer.

6. Which type of shoe got 1 fewer vote than [sneaker] ?
   Circle your answer.

7. Which type of shoe got the fewest votes?
   Circle your answer.

Name ______________________________

Range and Mode

Use [cube] to show each number.

Then find the mode and range.

1. 5, 1, 1, 1, 6 mode ___1___ range ___5___

2. 8, 6, 1, 3, 3 mode ______ range ______

3. 9, 3, 6, 4, 6 mode ______ range ______

4. 7, 5, 5, 5, 5 mode ______ range ______

5. 1, 2, 4, 3, 2 mode ______ range ______

6. 7, 6, 4, 6, 3 mode ______ range ______

7. 1, 3, 8, 4, 3 mode ______ range ______

8. 2, 2, 2, 4, 5 mode ______ range ______

9. 4, 3, 2, 1, 4 mode ______ range ______

10. 6, 1, 3, 5, 6 mode ______ range ______

Name ______________________________

Problem Solving: Strategy

12-5 PRACTICE

Make a Graph

Make a bar graph to solve.

1. The children go to the zoo. They see 2 tigers. They see 3 more lions than tigers. They see 1 more leopard than lions. They see the most of which big cat?

They see more ______________ than any other big cat.

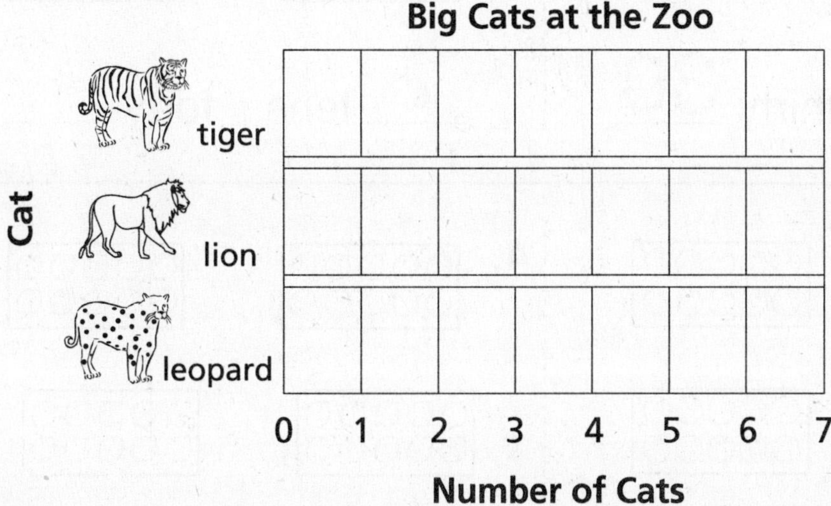

2. The aquarium has many sea animals. There are 6 sharks. There are 4 fewer dolphins than sharks. There are 3 more seals than dolphins. How many more sharks than seals are there? _______ more

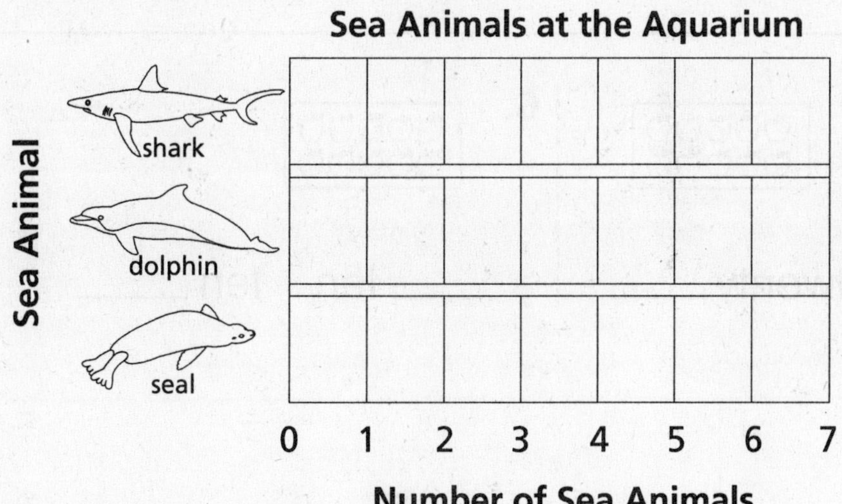

Name ______________________________

Tens

Count groups of ten.

Write each number.

1.  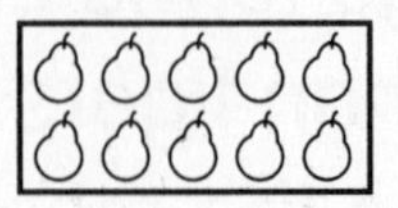

3 tens thirty 30

2. 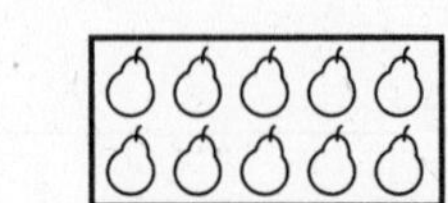

____ tens forty ____

3. 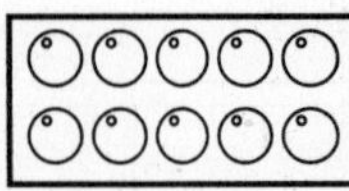 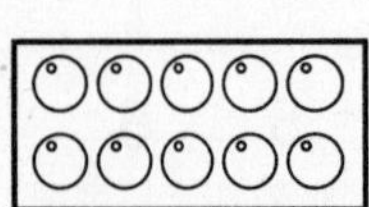

____ tens sixty ____

4. 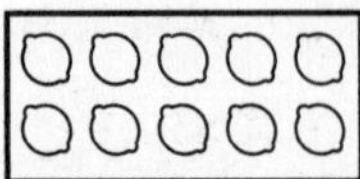 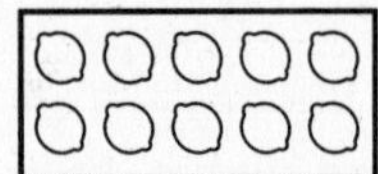 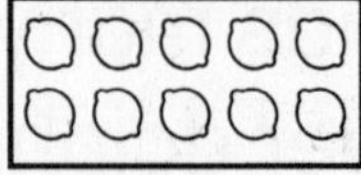

____ tens fifty ____

5.  

____ tens twenty ____

6. 

____ ten ten ____

Name ________________________________

Tens and Ones

Use [cube] to show each number.

Then make groups of tens and ones.

Write how many.

1. **26** twenty-six

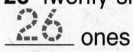

   __26__ ones

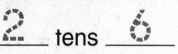

   __2__ tens __6__ ones

2. **31** thirty-one

   ______ ones

   ______ tens ______ one

3. **22** twenty-two

   ______ ones

   ______ tens ______ ones

4. **13** thirteen

   ______ ones

   ______ ten ______ ones

5. **24** twenty-four

   ______ ones

   ______ tens ______ ones

6. **45** forty-five

   ______ ones

   ______ tens ______ ones

7. **29** twenty-nine

   ______ ones

   ______ tens ______ ones

8. **38** thirty-eight

   ______ ones

   ______ tens ______ ones

Name ____________________

Numbers Through 100

Write the number in different ways.

1.

tens	ones

tens	ones
5	4

54 fifty-four

2.

tens	ones

tens	ones
____	____

____ ninety-three

3.

tens	ones

tens	ones
____	____

____ seventy

4.

tens	ones

tens	ones
____	____

____ thirty-eight

Name ____________________

Estimate Numbers

Circle 10. Estimate. Then count.

1. Estimate 20

Count 23

---

2. Estimate _____

Count _____

---

3. Estimate _____

Count _____

---

4. Estimate _____

Count _____

Name ____________________

Patterns on the Hundred Chart • Algebra

Use the hundred chart. Find each number in the box below on the chart. Find 10 less. Find 10 more.

1	2	3	4	5	6	7	8	9	10
11	12	13	14	15	16	17	18	19	20
21	22	23	24	25	26	27	28	29	30
31	32	33	34	35	36	37	38	39	40
41	42	43	44	45	46	47	48	49	50
51	52	53	54	55	56	57	58	59	60
61	62	63	64	65	66	67	68	69	70
71	72	73	74	75	76	77	78	79	80
81	82	83	84	85	86	87	88	89	90
91	92	93	94	95	96	97	98	99	100

1. **25** 10 less is 15 10 more is 35
2. **46** 10 less is ________ 10 more is ________
3. **90** 10 less is ________ 10 more is ________
4. **72** 10 less is ________ 10 more is ________
5. **64** 10 less is ________ 10 more is ________
6. **33** 10 less is ________ 10 more is ________
7. **59** 10 less is ________ 10 more is ________
8. **87** 10 less is ________ 10 more is ________

Use with Grade 1, Chapter 13, Lesson 6, pages 227–228.

Name ____________________

Problem Solving Skill: Reading for Math

Compare and Contrast

Maria goes to the beach.
She finds 36 and 1 .
Andy goes to the beach too.
He finds 51 and 1 .

1. What did Maria and Andy find that was the same? What did they find that was different?

____

____

2. Who found more shells?

____

3. Together, Maria and Andy found 87 shells. Write the number in tens and ones.

_______ tens _______ ones

Name ____________________

Compare Numbers to 100

Compare the numbers. Use tens and ones.

Compare. Write $>$, $<$, or $=$.

1.	72	(=)	72	**2.**	63	◯	76
3.	39	◯	40	**4.**	43	◯	34
5.	86	◯	88	**6.**	17	◯	18
7.	54	◯	45	**8.**	82	◯	82
9.	100	◯	98	**10.**	74	◯	94

Problem Solving

Solve.

Show Your Work

11. Which is true about 6 tens 5 ones?

The amount is greater than 68.

The amount is equal to 56.

The amount is less than 66.

12. Which is true about 3 tens 7 ones?

The amount is greater than 35.

The amount is equal to 39.

The amount is less than 28.

Name ______________________________

Order Numbers to 100

Write the number that comes just before and just after.

1. ← 38 →

2. ← 46 →

______ ______

3. ← 69 →

______ ______

4. ← 25 →

______ ______

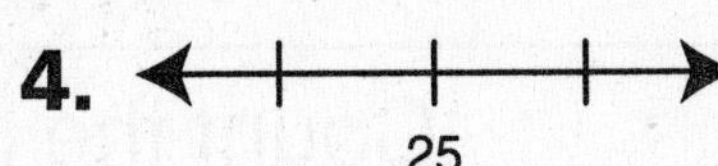

5. ← 90 →

______ ______

6. ← 33 →

______ ______

Write the number that is between.

7. ← 80 __ 82 →

______

8. ← 48 __ 50 →

______

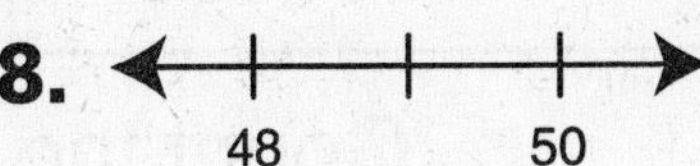

Problem Solving

Solve.

9. Jim wants the number after 29 on his shirt. Write the number on Jim's shirt.

Name ______________________________

Skip-Counting Patterns • Algebra

Skip count by twos, fives, or tens.

1. 

Count the leaves.

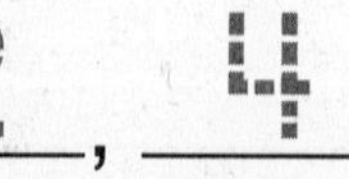

2, 4, 6, 8, 10 leaves

---

2. 

Count the leaves.

____, ____, ____, ____, ____, ____ leaves

---

3. 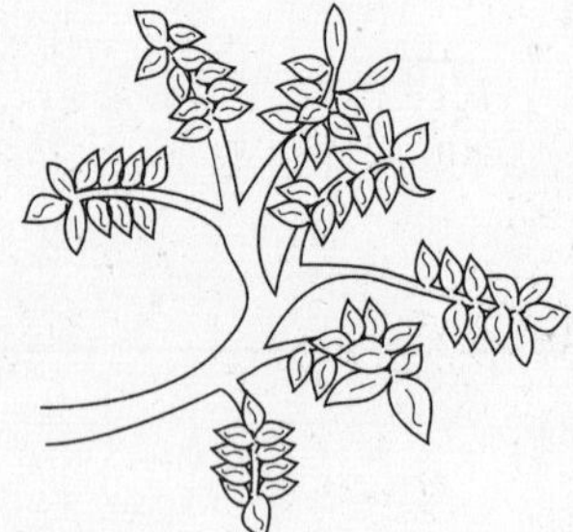

Count the leaves.

____, ____, ____, ____, ____, ____, ____ leaves

---

4. 

Count the leaves.

____, ____, ____, ____, ____, ____, ____ leaves

Make it Right

5. This is how Keith skip-counts by tens.

10, 20, 30, 50, 60, 70, 80

Why is Keith wrong? Make it right.

____

____

Name ____________________

Skip Count on a Hundred Chart • Algebra

1	2	3	4	5	6	7	8	9	10
11	12	13	14	15	16	17	18	19	20
21	22	23	24	25	26	27	28	29	30
31	32	33	34	35	36	37	38	39	40
41	42	43	44	45	46	47	48	49	50
51	52	53	54	55	56	57	58	59	60
61	62	63	64	65	66	67	68	69	70
71	72	73	74	75	76	77	78	79	80
81	82	83	84	85	86	87	88	89	90
91	92	93	94	95	96	97	98	99	100

1. Count by twos. Color the boxes with those numbers red.

2. Count by fives. Color the boxes with those numbers blue.

3. Count by tens. Put a box around those numbers.

4. Which numbers have 2 colors and a box?

____

Name ____________________

Even and Odd Numbers

Circle pairs. Write each number.

Circle odd or even.

1.  2

odd even

2.  ______

odd even

3. 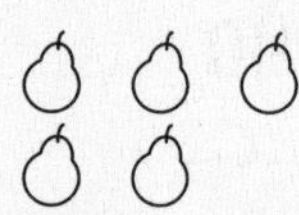 ______

odd even

4. 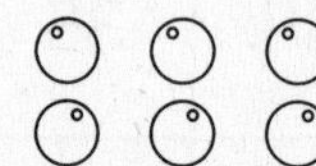 ______

odd even

5. 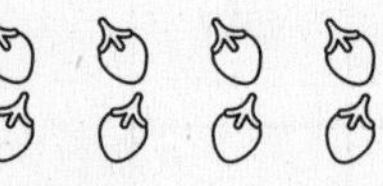 ______

odd even

6.  ______

odd even

Circle the even numbers.

7. 1, 2, 3, 4, 5, 6, 7, 8, 9, 10

8. 34, 35, 36, 37, 38, 39, 40, 41, 42, 43

Circle the odd numbers.

9. 11, 12, 13, 14, 15, 16, 17, 18, 19, 20

10. 20, 21, 22, 23, 24, 25, 26, 27, 28, 29

Name ______________________________

Problem Solving: Strategy

Make a Pattern

P 14-6 PRACTICE

Make a pattern. Solve.

1. Each bag has 5 apples. How many apples in 3 bags?

______ apples

Number of bags	Number of apples
1 bag	5
2 bags	______
3 bags	______

2. Each box has 2 melons. How many melons in 4 boxes?

______ melons

Number of boxes	Number of melons
1 box	______
2 boxes	______
3 boxes	______
4 boxes	______

3. Each sack has 10 oranges. How many oranges in 3 sacks?

______ oranges

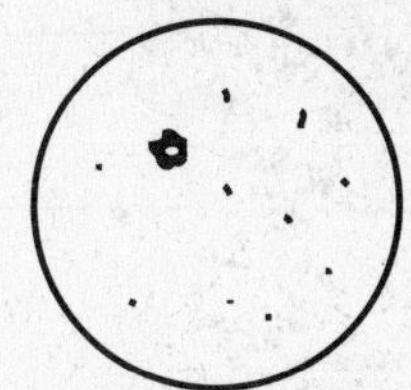

Number of sacks	Number of oranges
1 sack	______
2 sacks	______
3 sacks	______

Name ______________________________

Pennies and Nickels

You can use  and .

Count the coins to find each price. Write each price on the tag.

1.

5¢ 10¢ 11¢ 12¢ 13¢ 14¢

2.

____¢ ____¢ ____¢ ____¢ ____¢ ¢

3.

____¢ ____¢ ____¢ ____¢ ____¢ ¢

4.

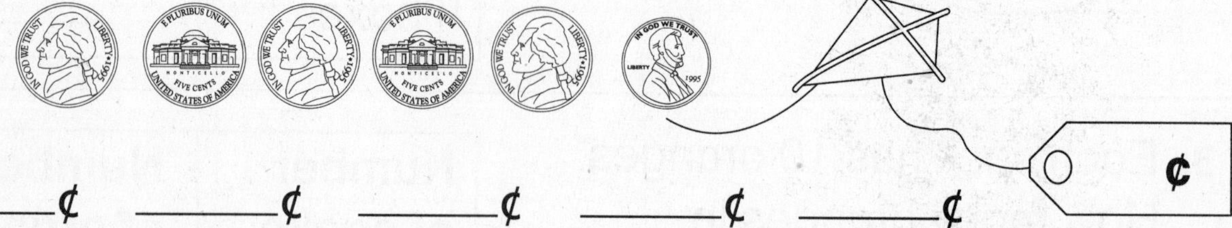

____¢ ____¢ ____¢ ____¢ ____¢ ¢

5.

____¢ ____¢ ____¢ ____¢ ____¢ ____¢ ¢

Name ______________________________

Pennies and Dimes

Trade for dimes.

Use  and  if you like.

Number of pennies	Trade for dimes	Amount
1. 36	10 10 10 1 1 1 1 1 1	36¢
2. 44		____¢
3. 51		____¢
4. 22		____¢
5. 13		____¢

Problem Solving

Solve.

6. Madison has 40¢ in pennies.
The machine only takes dimes.
What trade should she make?

____

Name ___

Pennies, Nickels, and Dimes

You can use coins.
Count to find each price.
Write each price on the tag.

1.      

10¢ 15¢ 20¢ 25¢ 26¢

2.      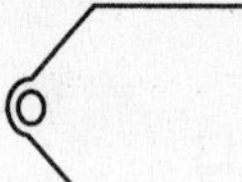

___¢ ___¢ ___¢ ___¢ ___¢

3.      

___¢ ___¢ ___¢ ___¢ ___¢

4.     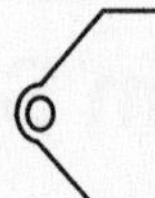

___¢ ___¢ ___¢ ___¢ ___¢

Use with Grade 1, Chapter 15, Lesson 3, pages 259–260.

Name ____________________

Counting Money

You can use coins.
Circle the coins to match each price.

1. 29¢

2. 33¢

3. 60¢

4. 45¢

5. 19¢

6. 30¢

Name ____________________

Problem Solving Skill: Reading for Math

Use Illustrations

Grandma takes Jeff to the beach.
He has money to buy one toy.
Jeff has 3 dimes.
He says, "I think I'll buy that pail."

Use the story and the picture.

1. How much money does Jeff have?

10¢ 20¢ 30¢

2. Which toy does Jeff want?

____

3. Can he buy it?

____

Name ______________________________

Equal Amounts

Use coins. Show each amount 2 ways.

Draw the coins you need.

1. 44¢

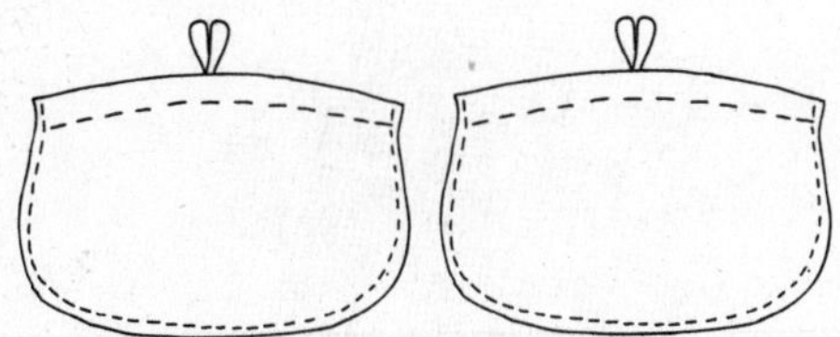

2. 35¢

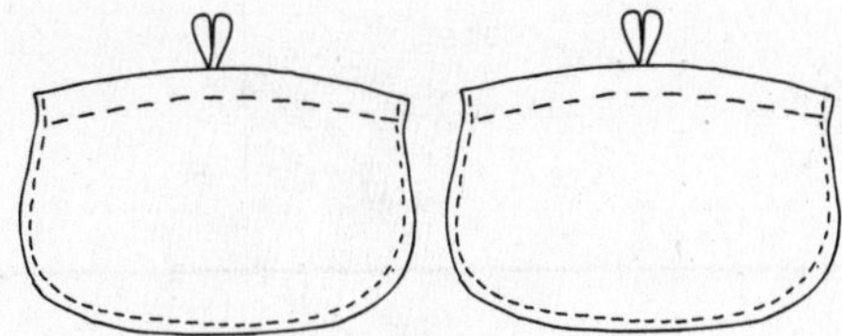

3. 41¢

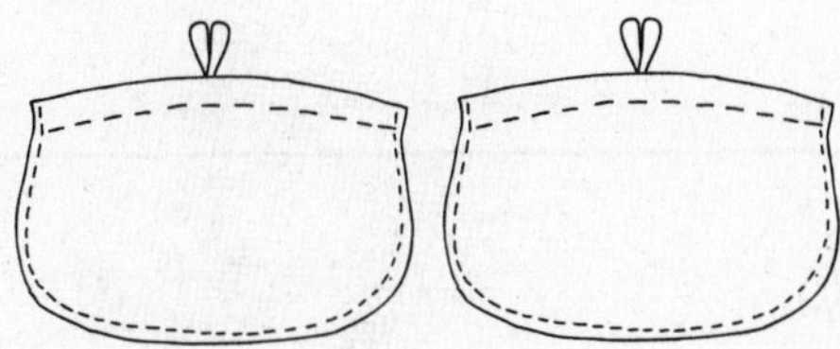

4. Maria has

Paula has

What coin does Paula need to have the same amount as Maria? ______

Name ____

Quarters

Use coins to make each amount.

Draw the coins you used.

1. Use (quarter) and (penny) to make 53¢.

25 25 1 1 1

2. Use (quarter) and (nickel) to make 40¢.

3. Use (quarter) and (dime) to make 70¢.

4. Use (quarter) and (dime) to make 55¢.

Make It Right

5. This is how Keith showed 67¢.

Why is Keith wrong?

____

____

Name ____________________

Dollar

You can use coins.
Make $1.00 in different ways.
Draw the coins.

Complete.

1. 

2.

3.

4.

Problem Solving

Solve.

5. Tracey has 4 dimes. Stacey has the same coins.

   Do they have more or less than $1.00 together?

   ______

6. Tanya has 6 dimes. Jarrette has the same coins.

   Do they have more or less than $1.00 together?

   ______

Name ______________________________

Money Amounts

Count the coins. Circle to tell if you can buy the toy.

1. 

Yes  No

2. 

Yes No

3. 

Yes No

Problem Solving

Solve.

Lori has 45¢. Her coins are all the same. They are not pennies. What coins does Lori have?

______

Show Your Work

Name ______________________________

Problem Solving: Strategy

Act It Out

Use coins to act out the problem. Solve.
Draw coins or write to explain.

1. Evan buys a for 17¢.

He buys a for 10¢.

How much money does Evan spend in all?

Evan spends ________¢.

---

2. Jane buys a  for 15¢.

She buys a for 20¢.

How much money does Jane spend in all?

Jane spends ________¢.

---

3. Frank buys a toy car for 32¢. Then he buys a ball for 20¢.

How much money does he spend in all?

He spends ________¢.

Name ____________________

Practice Addition Facts

17-1 PRACTICE

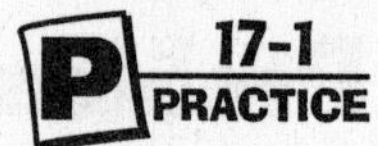

Add. Then color.

Sums of 11 or 12 — green

Sums of 9 or 10 — yellow

Sums of 7 or 8 — red

1 + 7
7 + 2
2 + 6
4 + 6
5 + 3
8 + 3
3 + 5
2 + 5
6 + 4
2 + 9
4 + 7
6 + 3
5 + 4
6 + 5
6 + 2
3 + 9
3 + 7
5 + 7
8 + 4
2 + 8
9 + 1
5 + 2
4 + 5
3 + 6
4 + 3
7 + 5
4 + 8
9 + 3
6 + 1
3 + 4

Use with Grade 1, Chapter 17, Lesson 1, pages 295–296.

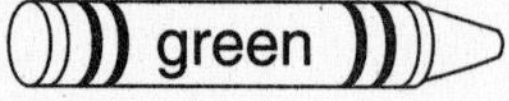

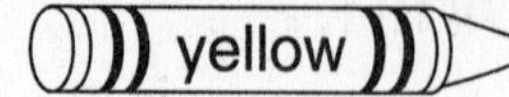

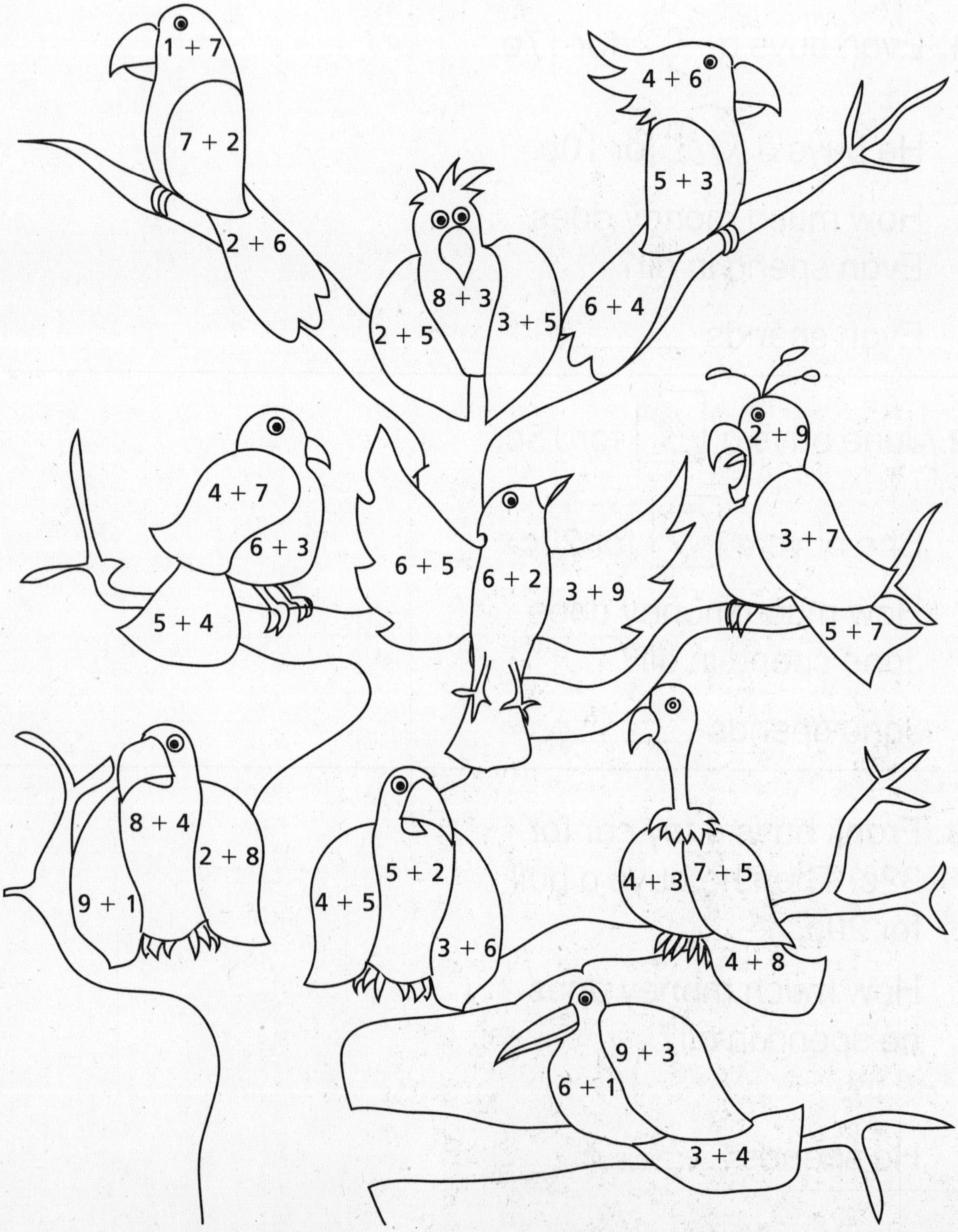

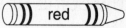

Name ______________________________

Doubles

Draw marbles to show the doubles.
Write the addends and the sums.

1. 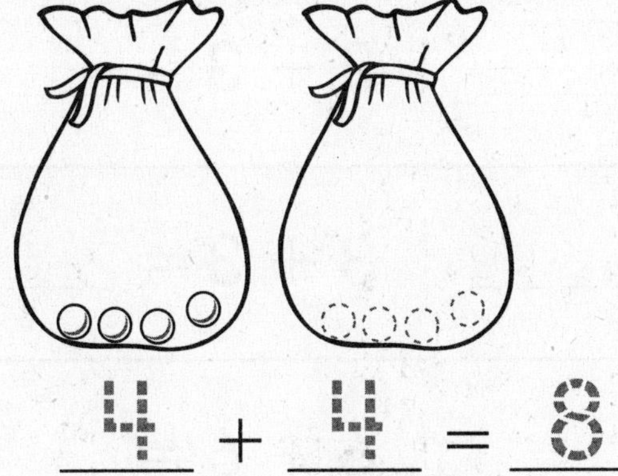

4 + 4 = 8

2. 

____ + ____ = ____

3. 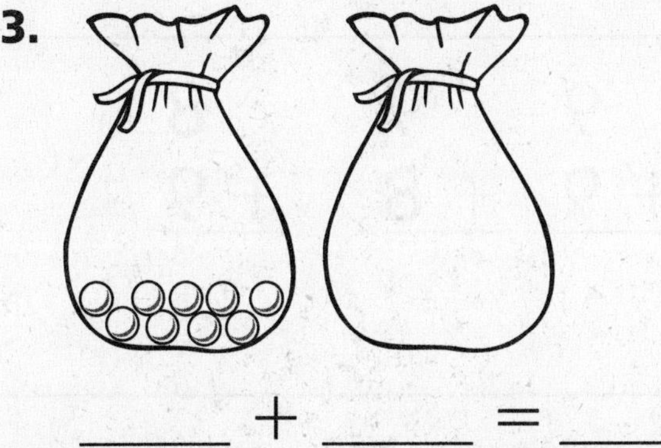

____ + ____ = ____

4. 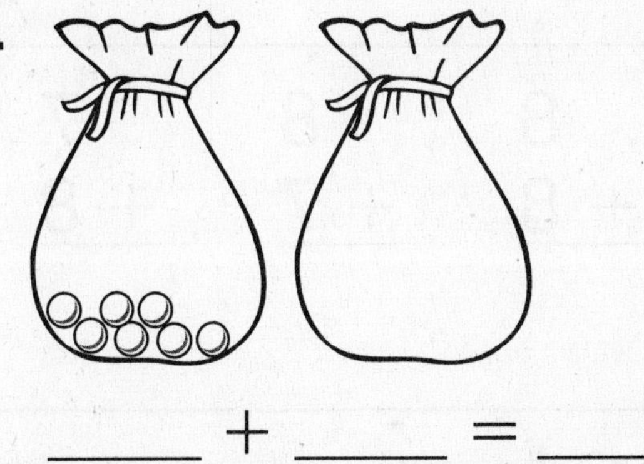

____ + ____ = ____

Add.

5. $\begin{array}{r} 6 \\ +\ 6 \\ \hline \end{array}$

6. $\begin{array}{r} 5 \\ +\ 5 \\ \hline \end{array}$

7. $\begin{array}{r} 8 \\ +\ 8 \\ \hline \end{array}$

8. $\begin{array}{r} 3 \\ +\ 3 \\ \hline \end{array}$

9. $\begin{array}{r} 2 \\ +\ 2 \\ \hline \end{array}$

Problem Solving

Solve.

10. Billy has 9 pennies. Wally has the same number of pennies. How many pennies do they have in all?

____ + ____ = ____ pennies

Show Your Work

Name ____

Doubles Plus 1

Find each sum. Use [cube].
Circle the doubles.

1. $2 + 2 =$ (4) | $5 + 5 =$ ____ | $2 + 3 =$ ____

2. $5 + 4 =$ ____ | $4 + 4 =$ ____ | $3 + 3 =$ ____

3. $6 + 7 =$ ____ | $7 + 7 =$ ____ | $6 + 6 =$ ____

4. $\begin{array}{r} 8 \\ +\,8 \\ \hline \end{array}$ $\begin{array}{r} 8 \\ +\,7 \\ \hline \end{array}$ $\begin{array}{r} 7 \\ +\,8 \\ \hline \end{array}$

5. $\begin{array}{r} 9 \\ +\,9 \\ \hline \end{array}$ $\begin{array}{r} 9 \\ +\,8 \\ \hline \end{array}$ $\begin{array}{r} 8 \\ +\,9 \\ \hline \end{array}$

Problem Solving

Solve.

Write a doubles plus 1 fact to solve.
What doubles fact can help you?

6. Lou Ann has 6 rubber bands.
Geri has 7 rubber bands.
How many rubber bands do they have in all?

$6 + 7 =$ ____ rubber bands

____ + ____ $= 12$ rubber bands

Show Your Work

Use with Grade 1, Chapter 17, Lesson 3, pages 299–300.

Name ____________________

Add Three Numbers • Algebra

Circle the numbers you add first.
Then find the sum.

1.	2.	3.	4.	5.
(6)	8	2	4	9
(4)	7	8	4	7
+ 7	+ 3	+ 5	+ 7	+ 1
17				

6.	7.	8.	9.	10.
5	8	7	6	3
5	2	7	5	6
+ 4	+ 6	+ 1	+ 4	+ 3

11.	12.	13.	14.	15.
2	6	9	8	5
3	6	6	1	7
+ 8	+ 1	+ 1	+ 8	+ 3

Problem Solving

Solve.

Choose the best strategy. Circle it.

16. 7 + 3 + 8 = _____ make a ten doubles

Name ____________________

Add Three Numbers in Any Order • Algebra

Find each sum.

1.

6	3	2
3	2	6
+ 2	+ 6	+ 3
11	11	11

2.

9	0	8
0	8	9
+ 8	+ 9	+ 0

3.

1	5	3
5	3	1
+ 3	+ 1	+ 5

4.

7	3	2
3	2	7
+ 2	+ 7	+ 3

5.

2	1	8
1	8	2
+ 8	+ 2	+ 1

6.

1	2	9
2	9	1
+ 9	+ 1	+ 2

Problem Solving

Solve.

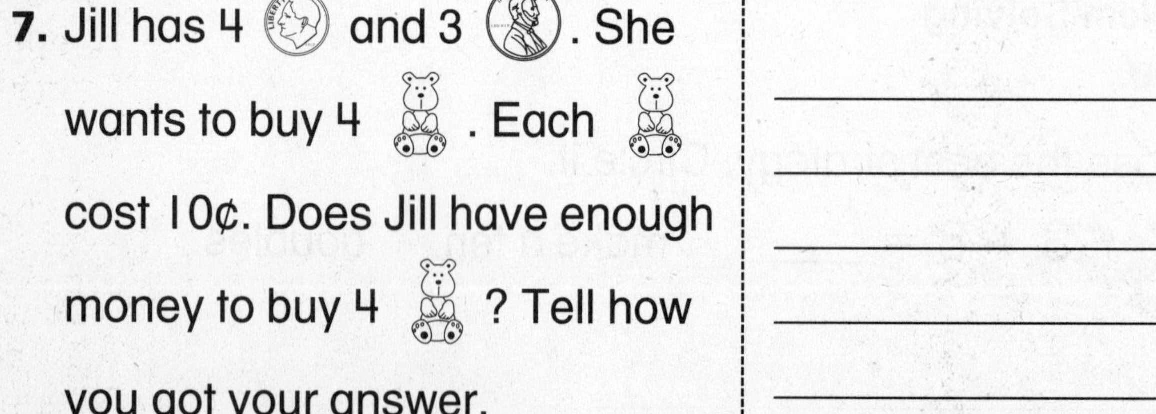

7. Jill has 4 (dimes) and 3 (pennies). She wants to buy 4 (bears). Each (bear) cost 10¢. Does Jill have enough money to buy 4 (bears)? Tell how you got your answer.

____

____

____

____

____

Name ______________________________

Problem Solving Skill: Reading for Math

Sequence of Events

Donna and her mom went to the pet store.
They saw many puppies.
First they saw 4 poodles. Next they saw 6 pugs.
Then 2 collies came over and licked them.

Sequence of Events

1. What was the last thing that happened?

____

2. How many poodles and pugs did Donna and her mom see? Write a number sentence.

3. How many puppies did they see in all?

4. Which puppies did they see first?

____

Name ____________________

Practice Subtraction Facts

Subtract.

Color the differences 0 to 5 yellow.

Color the differences 6 to 9 red.

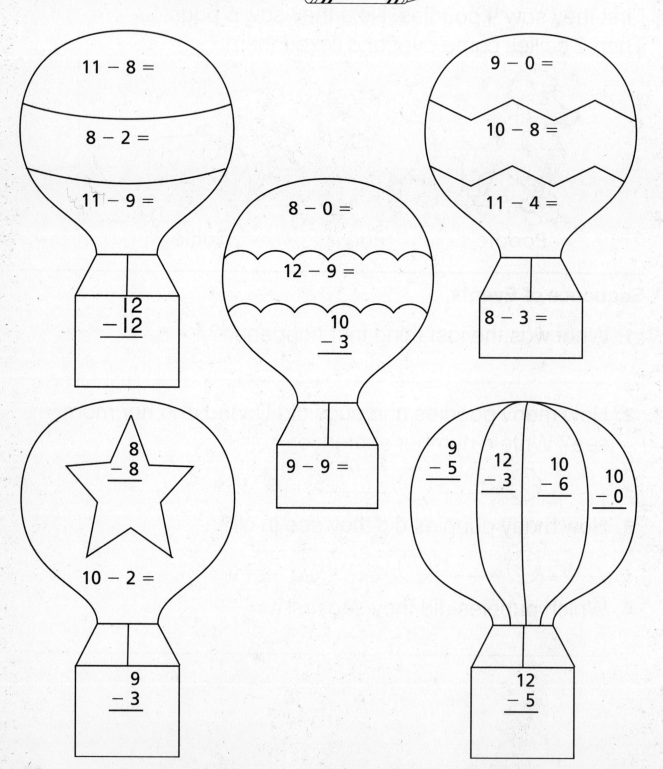

$11 - 8 =$

$8 - 2 =$

$11 - 9 =$

$12 - 12$

$9 - 0 =$

$10 - 8 =$

$11 - 4 =$

$8 - 3 =$

$8 - 0 =$

$12 - 9 =$

$10 - 3$

$9 - 9 =$

$8 - 8$

$10 - 2 =$

$9 - 3$

$9 - 5$

$12 - 3$

$10 - 6$

$10 - 0$

$12 - 5$

Name ______________________________

Use Doubles to Subtract

Add or subtract. Then draw a line to match the related facts.

1. 4 + 4 = 8	2 − 1 = ____
2. 7 + 7 = ____	14 − 7 = ____
3. 2 + 2 = ____	8 − 4 = ____
4. 1 + 1 = ____	4 − 2 = ____
5. 5 + 5 = ____	6 − 3 = ____
6. 9 + 9 = ____	16 − 8 = ____
7. 3 + 3 = ____	12 − 6 = ____
8. 8 + 8 = ____	10 − 5 = ____
9. 6 + 6 = ____	18 − 9 = ____

Name ____________________

Related Subtraction Facts

18-3 PRACTICE

Complete the related subtraction facts.

1. $11 - 7 = $ 4 $11 - 4 = $ 7	**2.** $10 - 8 = $ ___ $10 - 2 = $ ___	**3.** $12 - 5 = $ ___ $12 - 7 = $ ___
4. $9 - 4 = $ ___ $9 - 5 = $ ___	**5.** $12 - 9 = $ ___ $12 - 3 = $ ___	**6.** $11 - 6 = $ ___ $11 - 5 = $ ___
7. $11 - 9 = $ ___ $11 - 2 = $ ___	**8.** $9 - 3 = $ ___ $9 - 6 = $ ___	**9.** $17 - 8 = $ ___ $17 - 9 = $ ___
10. $8 - 5 = $ ___ $8 - 3 = $ ___	**11.** $10 - 7 = $ ___ $10 - 3 = $ ___	**12.** $8 - 6 = $ ___ $8 - 2 = $ ___
13. $10 - 8 = $ ___ $10 - 2 = $ ___	**14.** $11 - 3 = $ ___ $11 - 8 = $ ___	**15.** $13 - 6 = $ ___ $13 - 7 = $ ___
16. $15 - 8 = $ ___ $15 - 7 = $ ___	**17.** $12 - 4 = $ ___ $12 - 8 = $ ___	**18.** $7 - 3 = $ ___ $7 - 4 = $ ___

Use with Grade 1, Chapter 18, Lesson 3, pages 317–318.

Name ____________________

Relate Addition and Subtraction • Algebra

Use [cube]. Add. Then subtract.
Write the related subtraction fact.

1. $9 + 6 =$ 15

15 − 6 = 9

15 − 9 = 6

2. $4 + 8 =$ ___

___ ◯ ___ ◯ ___

___ ◯ ___ ◯ ___

3. $7 + 9 =$ ___

___ ◯ ___ ◯ ___

___ ◯ ___ ◯ ___

4. $8 + 5 =$ ___

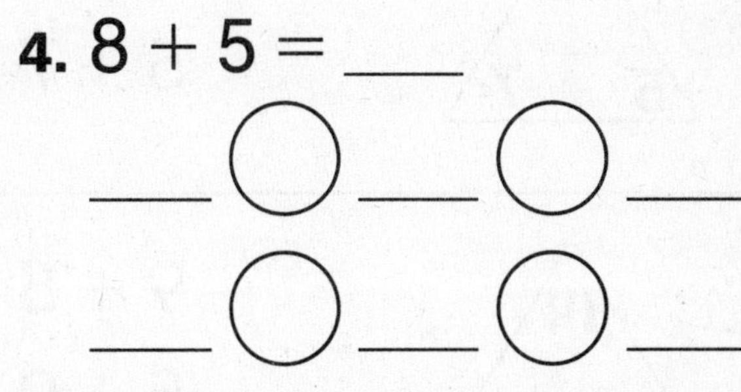

5. $8 + 3 =$ ___

___ ◯ ___ ◯ ___

___ ◯ ___ ◯ ___

6. $9 + 5 =$ ___

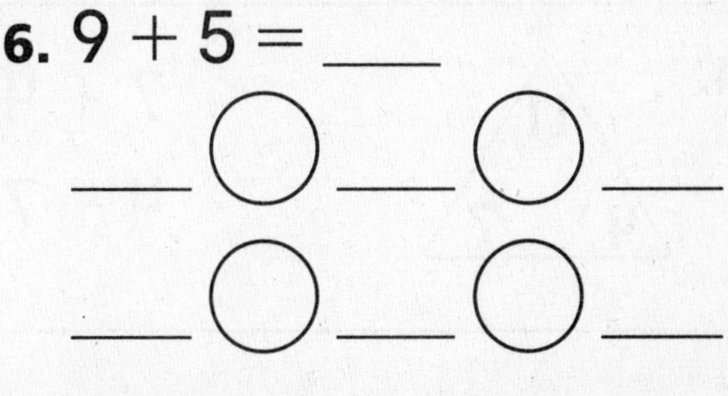

7. $9 + 4 =$ ___

___ ◯ ___ ◯ ___

___ ◯ ___ ◯ ___

8. $8 + 6 =$ ___

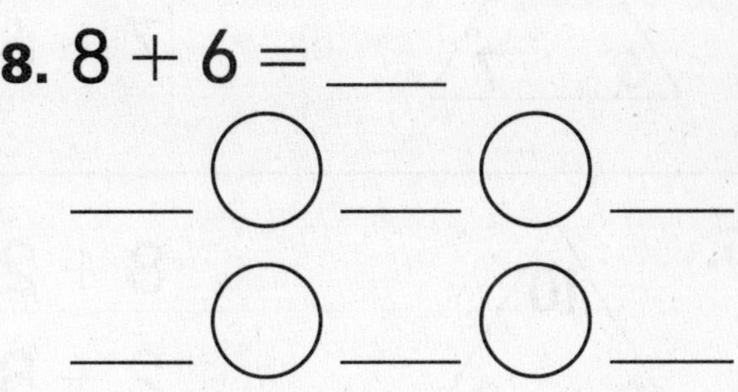

Name ______________________________

Fact Families

18-5 PRACTICE

Add and subtract.
Complete each fact family.

A fact family uses the same numbers.

1. 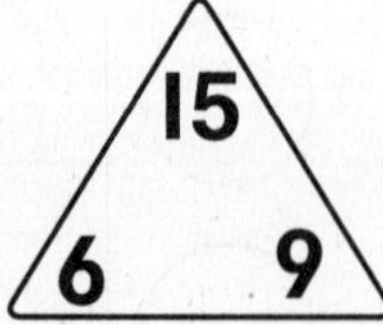

6 + 9 = 15 15 − 6 = 9
9 + 6 = ____ 15 − 9 = ____

2. 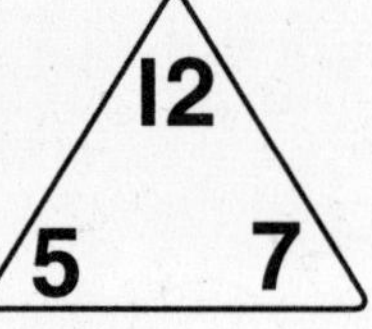

7 + 5 = ____ 12 − 5 = ____
5 + 7 = ____ 12 − 7 = ____

3. 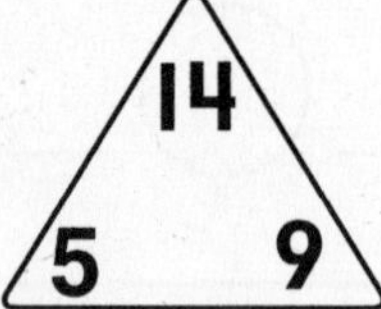

9 + 5 = ____ 14 − 5 = ____
5 + 9 = ____ 14 − 9 = ____

4. 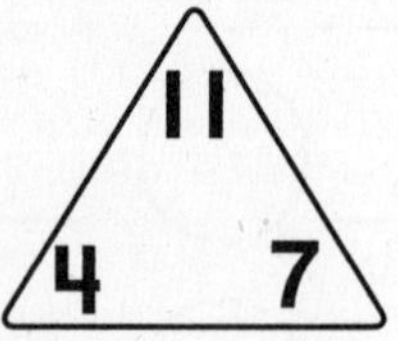

7 + 4 = ____ 11 − 4 = ____
4 + 7 = ____ 11 − 7 = ____

5. 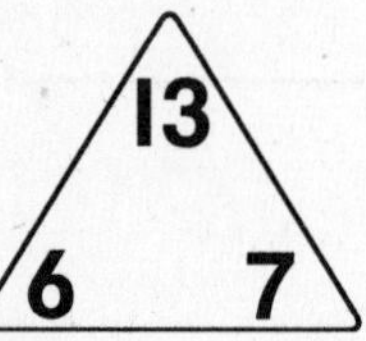

6 + 7 = ____ 13 − 7 = ____
7 + 6 = ____ 13 − 6 = ____

6. 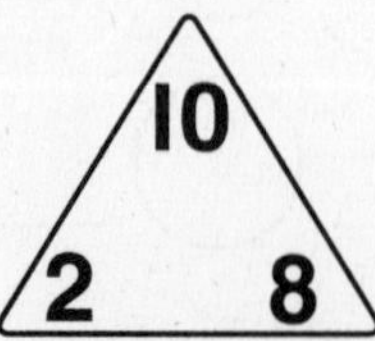

8 + 2 = ____ 10 − 2 = ____
2 + 8 = ____ 10 − 8 = ____

Use with Grade 1, Chapter 18, Lesson 5, pages 321–322.

Name ______________________________

Problem Solving: Strategy

Choose the Operation

P 18-6 PRACTICE

Circle *add* or *subtract*.
Write a number sentence to solve.

Show Your Work

1. 13 moths fly by the light. 8 fly away. How many moths are left?

+ add − subtract

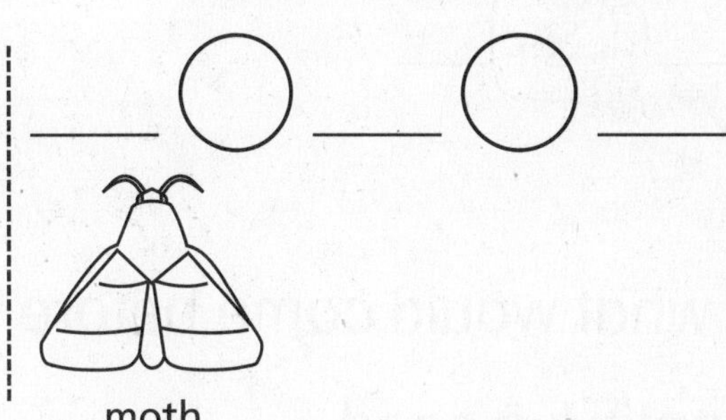

moth

2. 6 butterflies are in the garden. 5 more butterflies join them. How many butterflies are there in all?

+ add − subtract

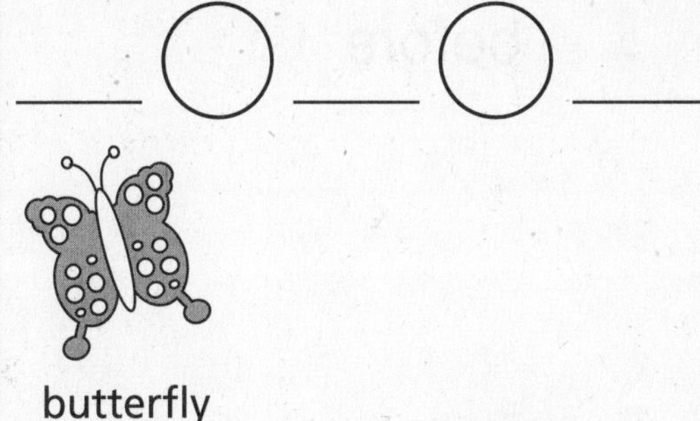

butterfly

3. Matt counts 7 inchworms on the leaves. He counts 4 inchworms on the flowers. How many inchworms are there in all?

+ add − subtract

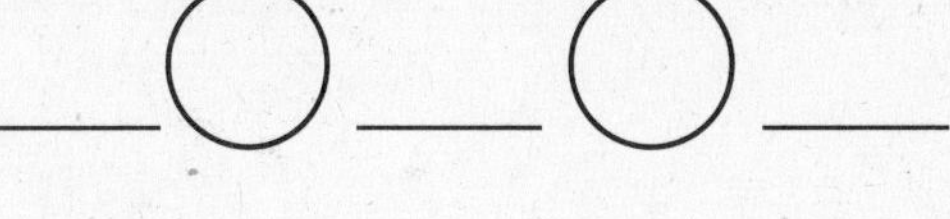

inchworm

4. 15 crickets chirp at night. 8 crickets stop chirping. How many crickets keep chirping?

+ add − subtract

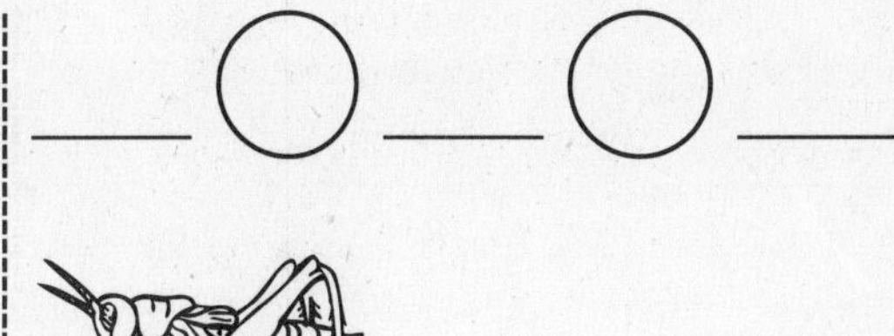

cricket

Name ____________________

Explore Time

Henry is building a house with blocks.

before

after

Draw what would come before and after.

1. Planting a seed.

before	after

2. Raking the leaves.

before	after

Name ____________________

Read the Clock

Use the  to help you answer the questions.

1. 

Where is the minute hand? ______

Where is the hour hand? ______

The time is ______ o'clock.

---

2. 

Where is the minute hand? ______

Where is the hour hand? ______

The time is ______ o'clock.

---

3. Draw the minute hand to point to the 12.
Draw the hour hand to point to the 4.
Your clock says 4 o'clock.

---

4. Draw the minute hand to point to the 12.
Draw the hour hand to point to the 10.
Your clock says 10 o'clock.

Make It Right

5. Ling says the minute hand is on the 2
and the hour hand is on the 12.
Why is Ling wrong?
Make it right.

Name ____

Time to the Hour

Use . Write the time.

1.

___3___ o'clock ______ o'clock ______ o'clock

2.

 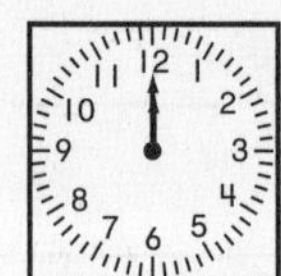 

______ o'clock ______ o'clock ______ o'clock

3.

 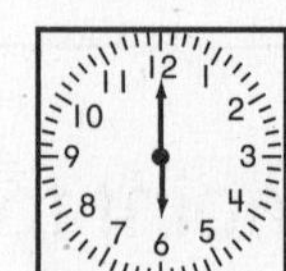 

______ o'clock ______ o'clock ______ o'clock

4.

______ o'clock ______ o'clock ______ o'clock

Name ______________________________

Time to the Half Hour

Use . Write the time.

1. 

half past ______

2. 

half past ______

3. 

half past ______

4. 

half past ______

5. 

half past ______

6. 

half past ______

Problem Solving

Solve. Read. Look at the clock. Write the time.

7. What time does Sami eat breakfast?

half past ______

8. What time does Sami eat lunch?

half past ______

9. What time does Sami eat dinner?

half past ______

Name ______________________________

Hour and Half Hour

Write each time. Circle the later time.

1.

2.

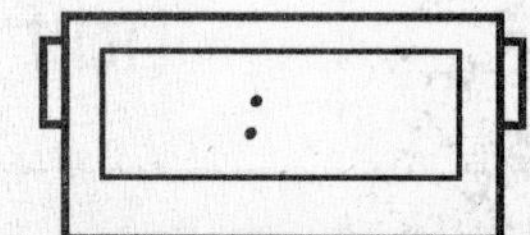

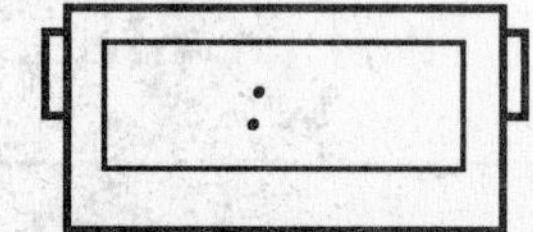

3.

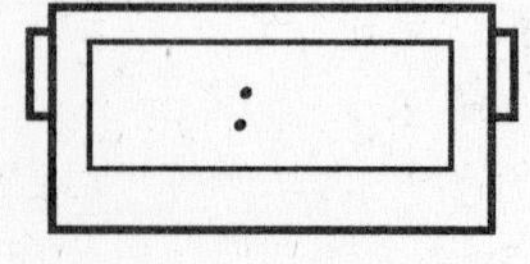

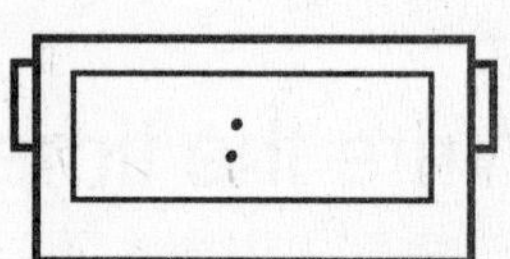

4.

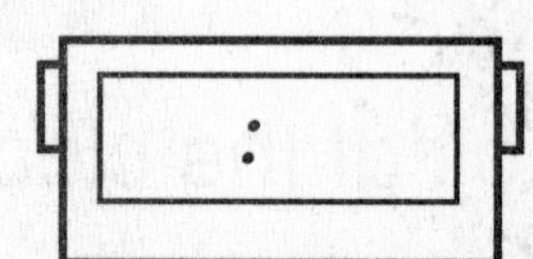

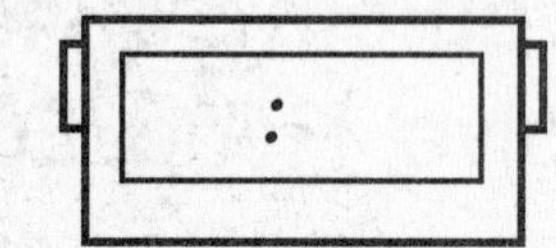

5.

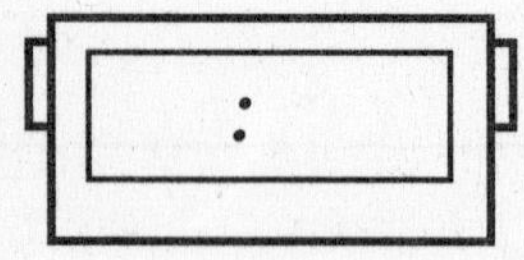

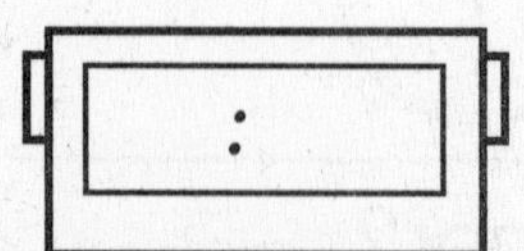

6.

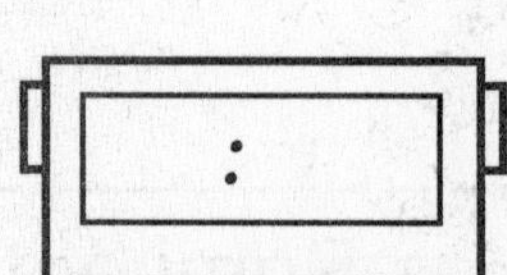

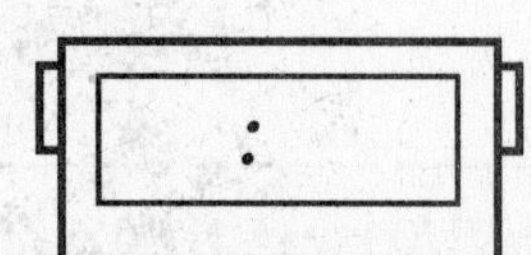

Name ______________________________

Practice Telling Time

Draw the hands.

1. 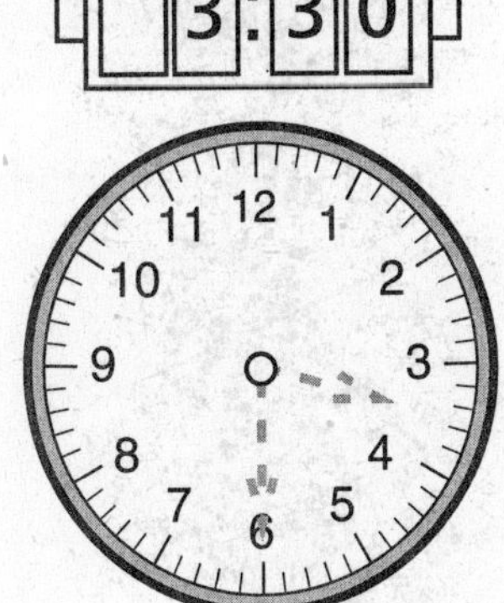

2. 

3. 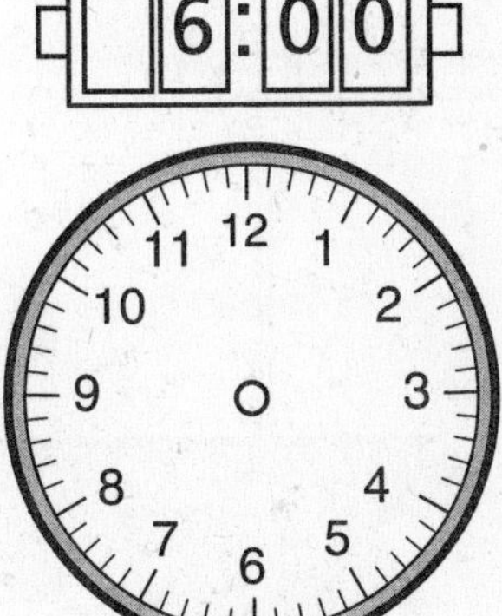

4. 

5. 

6. 

Problem Solving

Solve. Use [clock] to help.

7. Brian starts with this time. Brian moves the minute hand 60 minutes. What time is it now?

____

Name ____________________

Problem Solving Skill: Reading for Math

P 19-7 PRACTICE

Make Inferences

Olivia goes to the Play Center after school. She stays there until her mother picks her up. Olivia meets her friends at the center. Playing games is her favorite activity.

Play Center Activities	
Time	Activity
3:30	Snack
4:00	Homework
4:30	Movie
5:30	Games

Solve.

1. What time does Olivia start her homework?
   _______ o'clock
2. What time does the movie start? ___:___
3. How long does the movie last? ____________________
4. What time does Olivia's favorite activity begin? ___:___

Name ____________________

Estimate Time

Circle your answers.

Activity	About how long will it take?	How will you measure?

1. Sara sleeps from night till morning.

8 — minutes / hours / months

May						
S	M	T	W	T	F	S
				1	2	3
4	5	6	7	8	9	10
11	12	13	14	15	16	17
18	19	20	21	22	23	24
25	26	27	28	29	30	31

2. Jack watches a game show on TV.

30 — minutes / hours / days

June						
S	M	T	W	T	F	S
1	2	3	4	5	6	7
8	9	10	11	12	13	14
15	16	17	18	19	20	21
22	23	24	25	26	27	28
29	30					

3. Jamal becomes a teenager.

7 — hours / months / years

September						
S	M	T	W	T	F	S
	1	2	3	4	5	6
7	8	9	10	11	12	13
14	15	16	17	18	19	20
21	22	23	24	25	26	27
28	29	30				

4. Bao eats a bowl of cereal.

15 — minutes / hours / days

August						
S	M	T	W	T	F	S
					1	2
3	4	5	6	7	8	9
10	11	12	13	14	15	16
17	18	19	20	21	22	23
24	25	26	27	28	29	30
31						

5. Linda takes a vacation.

1 — minute / hour / week

December						
S	M	T	W	T	F	S
	1	2	3	4	5	6
7	8	9	10	11	12	13
14	15	16	17	18	19	20
21	22	23	24	25	26	27
28	29	30	31			

Name ______________________________

Use a Calendar

January

S	M	T	W	T	F	S
				1	2	3
4	5	6	7	8	9	10
11	12	13	14	15	16	17
18	19	20	21	22	23	24
25	26	27	28	29	30	31

February

S	M	T	W	T	F	S
1	2	3	4	5	6	7
8	9	10	11	12	13	14
15	16	17	18	19	20	21
22	23	24	25	26	27	28
29						

March

S	M	T	W	T	F	S
	1	2	3	4	5	6
7	8	9	10	11	12	13
14	15	16	17	18	19	20
21	22	23	24	25	26	27
28	29	30	31			

April

S	M	T	W	T	F	S
				1	2	3
4	5	6	7	8	9	10
11	12	13	14	15	16	17
18	19	20	21	22	23	24
25	26	27	28	29	30	

May

S	M	T	W	T	F	S
						1
2	3	4	5	6	7	8
9	10	11	12	13	14	15
16	17	18	19	20	21	22
23	24	25	26	27	28	29
30	31					

June

S	M	T	W	T	F	S
		1	2	3	4	5
6	7	8	9	10	11	12
13	14	15	16	17	18	19
20	21	22	23	24	25	26
27	28	29	30			

July

S	M	T	W	T	F	S
				1	2	3
4	5	6	7	8	9	10
11	12	13	14	15	16	17
18	19	20	21	22	23	24
25	26	27	28	29	30	31

August

S	M	T	W	T	F	S
1	2	3	4	5	6	7
8	9	10	11	12	13	14
15	16	17	18	19	20	21
22	23	24	25	26	27	28
29	30	31				

September

S	M	T	W	T	F	S
			1	2	3	4
5	6	7	8	9	10	11
12	13	14	15	16	17	18
19	20	21	22	23	24	25
26	27	28	29	30		

October

S	M	T	W	T	F	S
					1	23
3	4	5	6	7	8	9
10	11	12	13	14	15	16
17	18	19	20	21	22	23
24	25	26	27	28	29	30
31						

November

S	M	T	W	T	F	S
	1	2	3	4	5	6
7	8	9	10	11	12	13
14	15	16	17	18	19	20
21	22	23	24	25	26	27
28	29	30				

December

S	M	T	W	T	F	S
			1	2	3	4
5	6	7	8	9	10	11
12	13	14	15	16	17	18
19	20	21	22	23	24	25
26	27	28	29	30	31	

1. Which is the fifth month?

____

2. How many Thursdays are there in April?

____

3. Which month comes just before March?

____

4. How many days are there in January?

____

5. Which month comes just after May?

____

Name ____________________

Problem Solving: Strategy
Find a Pattern

Find the patterns to answer the questions.

June

S	M	T	W	T	F	S
					3	4
		7	8			11
12			15			
19	20				24	
	27	28	29			

1. Where are the flowers on the calendar?

____

How do you know?

____

2. Look at the smiley faces. Which dates would the next two smiley faces fall on? Write the dates. Put faces on the calender.

____

How do you know?

____

3. How are June 9 and June 30 alike?

____

How do you know?

____

Name ______________________________

Explore Length

21-1 PRACTICE

Estimate how many ▢ long.

Then use ▢ to measure.

1. 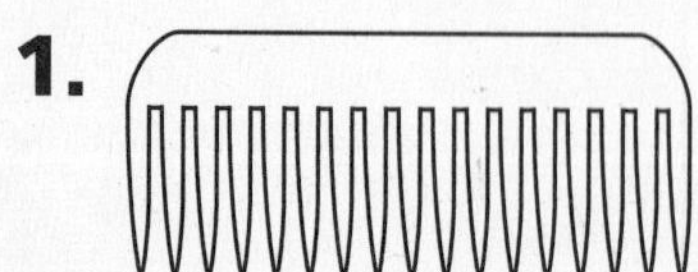

Estimate: about ______ ▢ long

Measure: about ______ ▢ long

2. 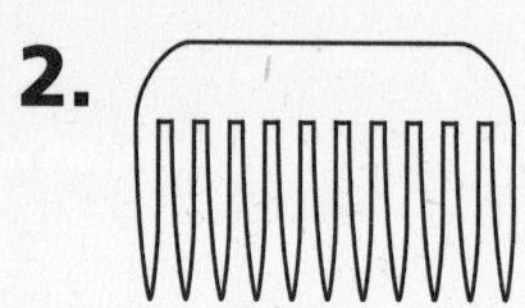

Estimate: about ______ ▢ long

Measure: about ______ ▢ long

3. 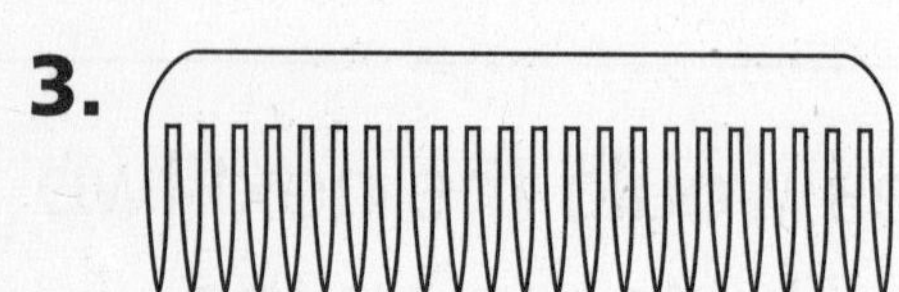

Estimate: about ______ ▢ long

Measure: about ______ ▢ long

4. 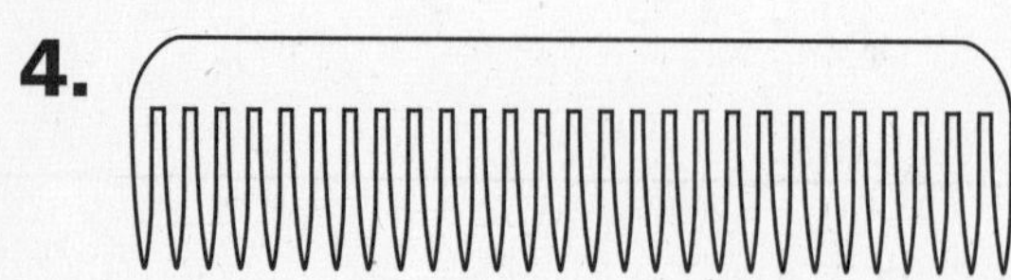

Estimate: about ______ ▢ long

Measure: about ______ ▢ long

Name ______________________________

Inch

Find these objects in your classroom.
Estimate how long.
Then use an inch ruler to measure.

1.

Estimate: about ______ inches

Measure: about ______ inches

2.

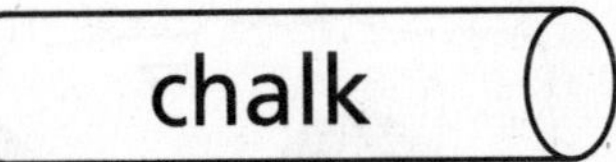

Estimate: about ______ inches

Measure: about ______ inches

3.

Estimate: about ______ inches

Measure: about ______ inches

4.

Estimate: about ______ inches

Measure: about ______ inches

5.

Estimate: about ______ inches

Measure: about ______ inches

6.

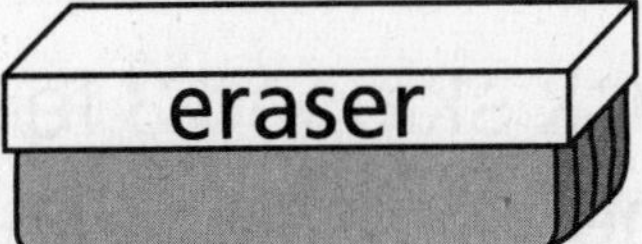

Estimate: about ______ inches

Measure: about ______ inches

Name ____________________

Inch and Foot

Which is better for measuring the real object?
Circle inches or feet.

1.

inches feet

2.

inches feet

3.

inches feet

4.

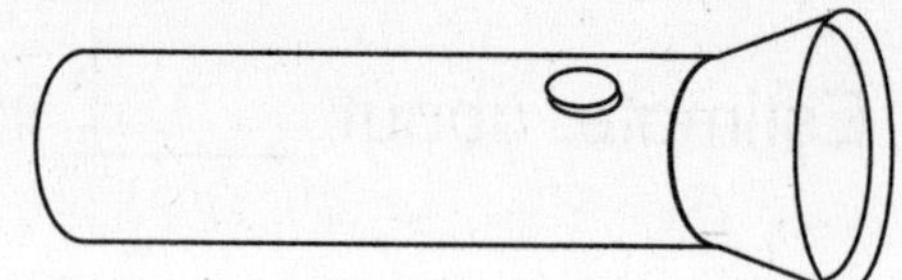

inches feet

5.

inches feet

6.

inches feet

Problem Solving

Solve.

7. Jack skipped 3 feet.
Then he skipped 4 feet.
How many feet did he skip in all?

_____ feet

Show Your Work

Name ______________________

Understanding Measurement

Connect the dots in order. Use an inch ruler to draw and measure the lines.

1. How long is the line from A to B?

   __4__ inches

2. How long is the line from B to C?

   ______ inches

3. How long is the line from C to D?

   ______ inches

4. Look at the line from D to A. About how long is it?

   Estimate: about ______ inches

   Measure: about ______ inches

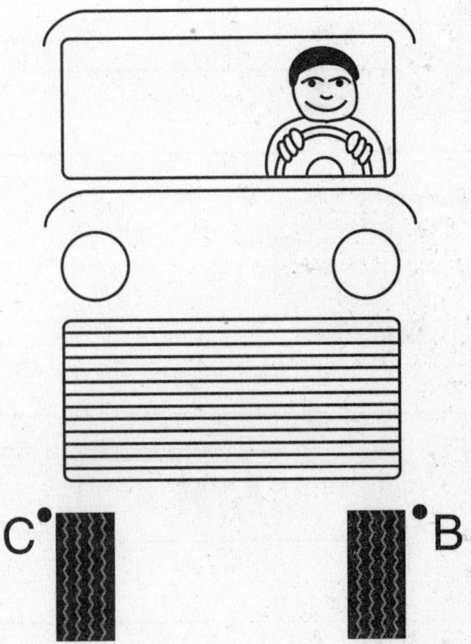

Problem Solving

Solve. Estimate.

5. The bottom of this shape is 2 inches long. About how long are the other sides? Estimate.

   From A to B: about ______ inches

   From C to B: about ______ inches

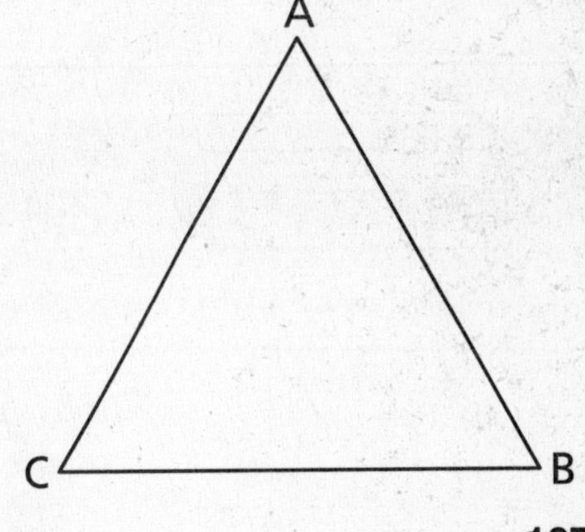

Name ____________________

Centimeter

Find these things in your classroom. Estimate how long. Then use a centimeter ruler to measure.

	Find	Estimate	Measure
1.	red	about _____ centimeters	about _____ centimeters
2.		about _____ centimeters	about _____ centimeters
3.		about _____ centimeters	about _____ centimeters
4.	MATH	about _____ centimeters	about _____ centimeters
5.		about _____ centimeters	about _____ centimeters
6.		about _____ centimeters	about _____ centimeters

Use with Grade 1, Chapter 21, Lesson 5, pages 379–380.

Name ____

Problem Solving Skill: Reading for Math

Predict Outcomes

Rob and Lizzie like to find things to measure. They each make handprints in the snow. Dad says, "I can measure the handprints with my tape measure." Rob's handprint is 4 inches long. Lizzie's handprint is 5 inches long. Then mom and dad make handprints in the snow.

Complete.

1. How much longer is Lizzie's handprint than Rob's?
   _______ inch

2. Dad measures his handprint. It is 9 inches long. Dad's handprint is _______ inches longer than Lizzie's.

3. Do you think Mom's handprint is longer or shorter than Lizzie's? _________

4. How many inches long do you think Mom's handprint is?
   _______ inches

5. What do you think Lizzie and Rob will measure next?

   ____

Name ______________________________

Explore Weight

Compare each object to a .

Circle your estimate. Then use a to measure. Circle your answer.

	Object	Estimate	Measure
1.		heavier lighter	heavier lighter
2.		heavier lighter	heavier lighter
3.		heavier lighter	heavier lighter

Problem Solving

Solve.

Circle the heavier object.

4.

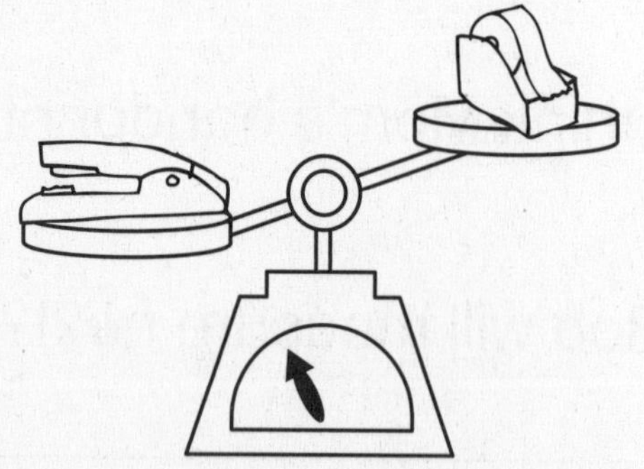

5.

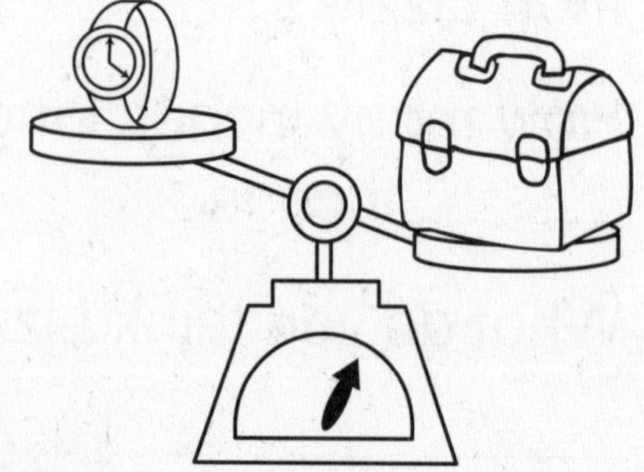

Name ____

Cup, Pint, Quart

Compare each container to the 1 cup, 1 pint, or 1 quart.
Circle the best estimate.

		Container	Estimate
1.	1 quart		more than 1 quart less than 1 quart
2.	1 cup	PAINT	more than 1 cup less than 1 cup
3.	1 pint	JUICE	more than 1 pint less than 1 pint
4.	quart	Water	more than 1 quart less than 1 quart
5.	1 cup	Teaspoon	more than 1 cup less than 1 cup

Name ____________________

Pound

Estimate. Circle your answer.

Then measure with  and a .

	Object	Estimate	Measure
1.		less than 1 pound more than 1 pound	less than 1 pound more than 1 pound
2.		less than 1 pound more than 1 pound	less than 1 pound more than 1 pound
3.		less than 1 pound more than 1 pound	less than 1 pound more than 1 pound
4.		less than 1 pound more than 1 pound	less than 1 pound more than 1 pound
5.		less than 1 pound more than 1 pound	less than 1 pound more than 1 pound

Name ____________________

Liter

Estimate.
If the container holds less than 1 liter, write the word less.
If it holds more than 1 liter, write the word more.

1. 

____

2. 

____

3. 

____

4. 

____

5. 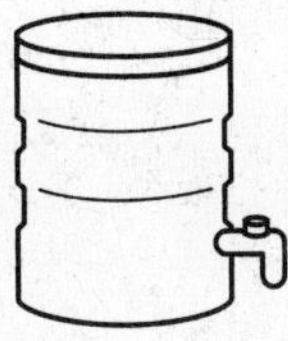

____

6. 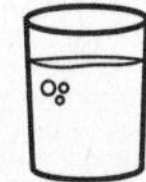

____

7. 

____

8. 

____

9. 

____

Problem Solving

Solve. Use the table.

10. Mom makes punch for a party. How much grape juice does she use?

____

11. Of which kind of juice does Mom use the most?

____

Mom's Juice Punch	
Apple	2 liters
Grape	1 liter
Orange	3 liters

Name ____________________

Gram and Kilogram

Draw a line from the object to the unit of measure you would use.

1. 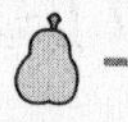 gram

 kilogram

2.  gram

kilogram

3.  gram

kilogram

4.  gram

 kilogram

5.  gram

 kilogram

6.   gram

kilogram

Problem Solving

Circle your answer.

7. How heavy is a small rabbit?

about 3 gram

about 3 kilograms

8. How heavy is a peanut?

about 3 grams

about 3 kilograms

Name ____

Temperature

Write the temperature.

1.

30 °F

2.

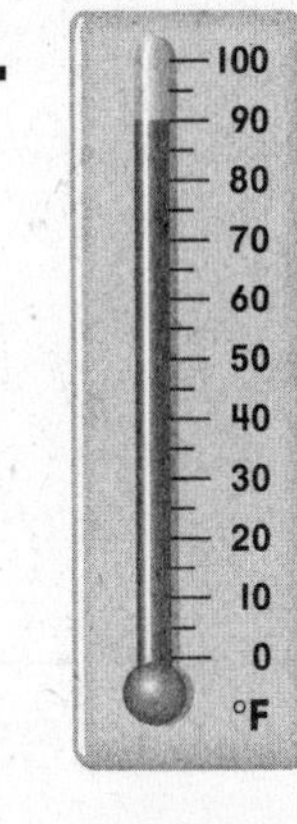

_____ °F

3.

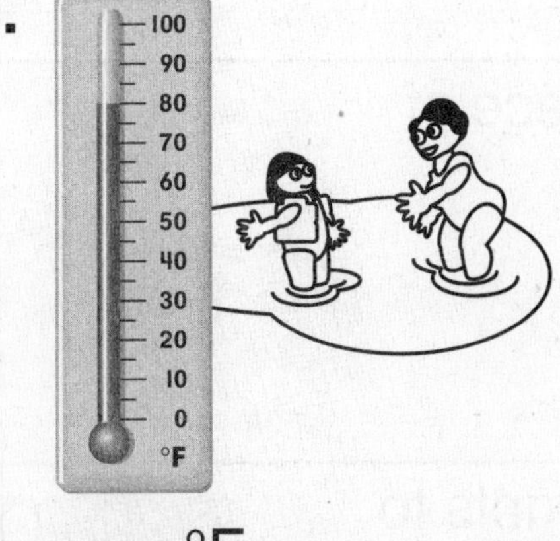

_____ °F

4.

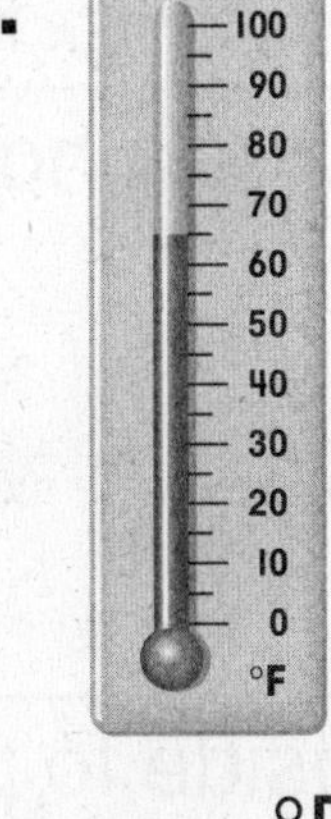

_____ °F

5.

_____ °C

6.

_____ °C

Name ___

Problem Solving: Strategy

Use Logical Reasoning

Complete. Circle the measuring tool you would use.

1. Mario caught a fish. He wants to know how much it weighs.

 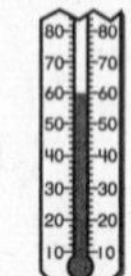  

fish

2. Karen is making cookies. She wants to measure the flour.

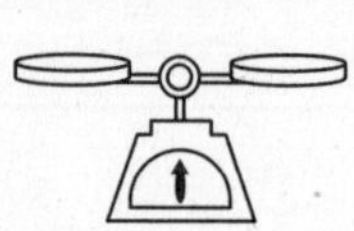 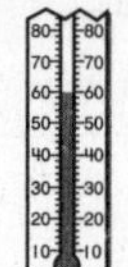  

3. Henry wants to know how long a piece of string is.

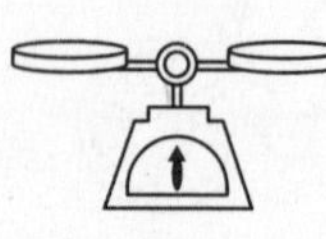 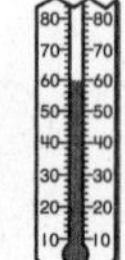  

string

4. Lisa is going outside to play. She wants to know how cold it is.

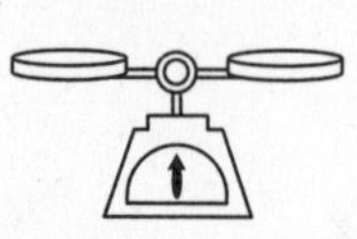 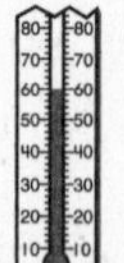  

outside

5. Sharon has a pail filled with sand. She wants to know how much it weighs.

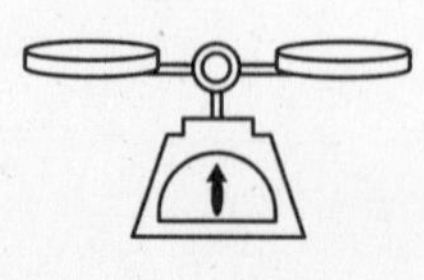 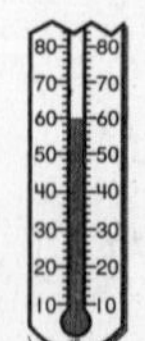  

Name ____________________

3-Dimensional Figures

Color the shapes.

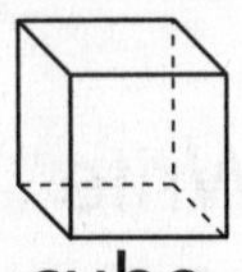

cube — red

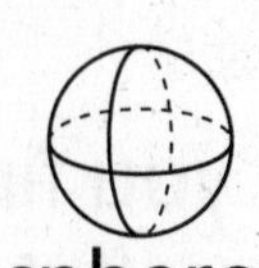

sphere — blue

cone — yellow

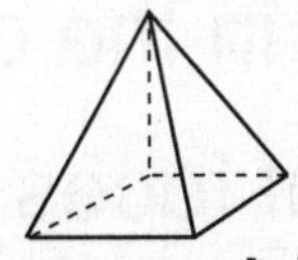

pyramid — green

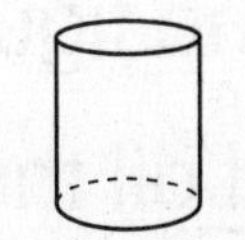

cylinder — purple

rectangular prism — orange

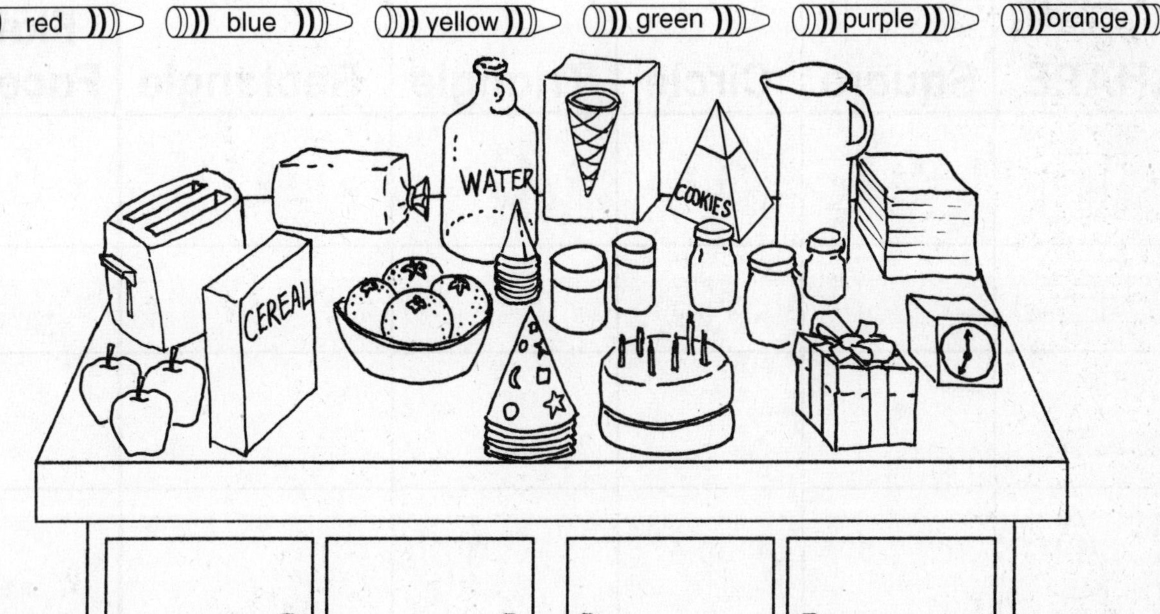

Problem Solving

Solve.

Sort the objects into two groups. Circle each object in one group. Underline each object in the other group.

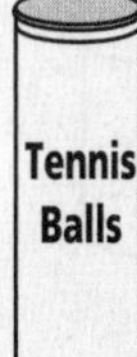

Explain your sorting rule. ______________________________

____

Name ____________________

2- and 3-Dimensional Figures

Find objects in your classroom. The objects should be shaped like the figures in the chart.

Trace around all the flat faces of the objects you find. Write how many shapes you found. Write how many flat faces.

SHAPE	Square	Circle	Triangle	Rectangle	Flat Faces
1.	6	0	0	0	6
2.					
3.					
4.					
5.					

Problem Solving

Solve.

Color 2-dimensional objects red.

Color 3-dimensional objects blue.

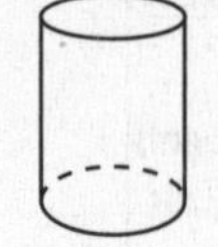
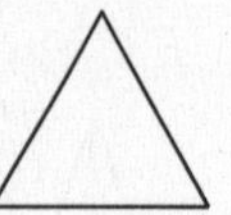

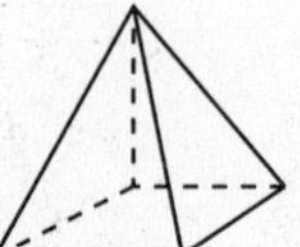

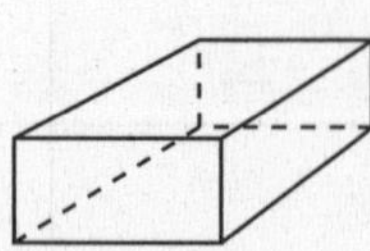
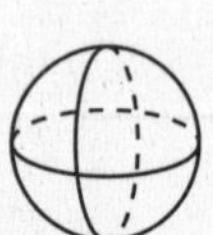
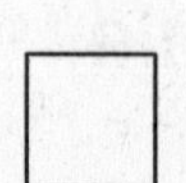
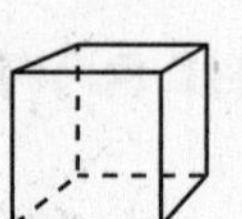

Use with Grade 1, Chapter 23, Lesson 2, pages 411–412.

Name ____________________

Build Shapes

Use pattern blocks to make each shape. Draw how you made it.

Make this shape.	Use this block.	Draw the shape.
1.		
2.		
3.		
4.		

Name ____

Sides and Vertices

Color all the shapes that belong with each rule.

1. 3 sides

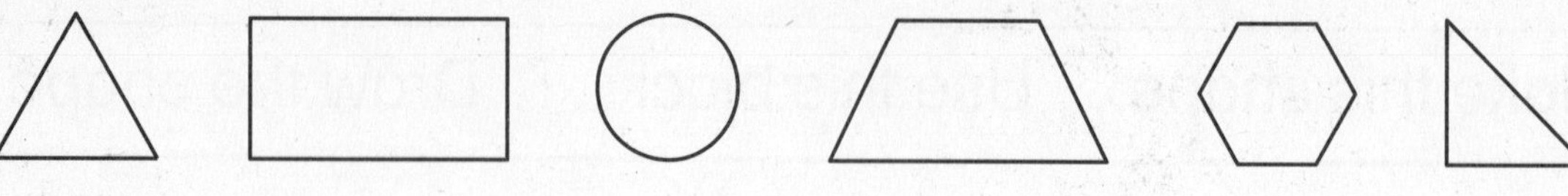

2. 4 sides

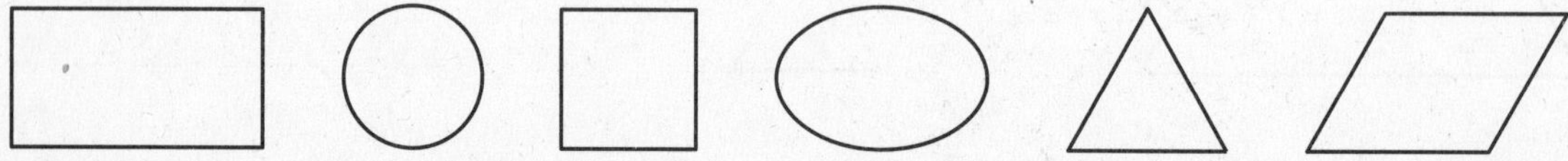

3. more than 4 vertices

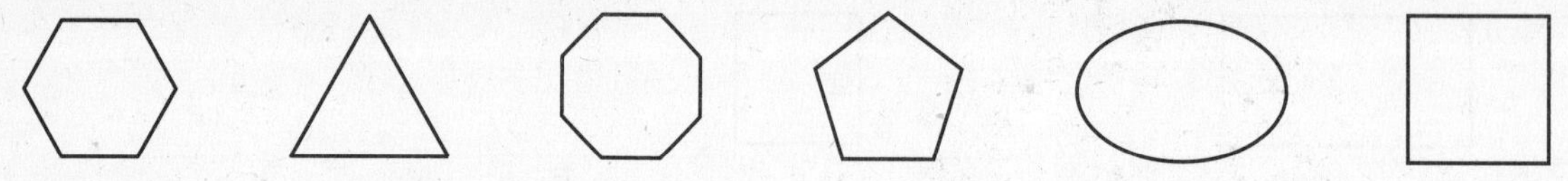

4. fewer than 4 vertices

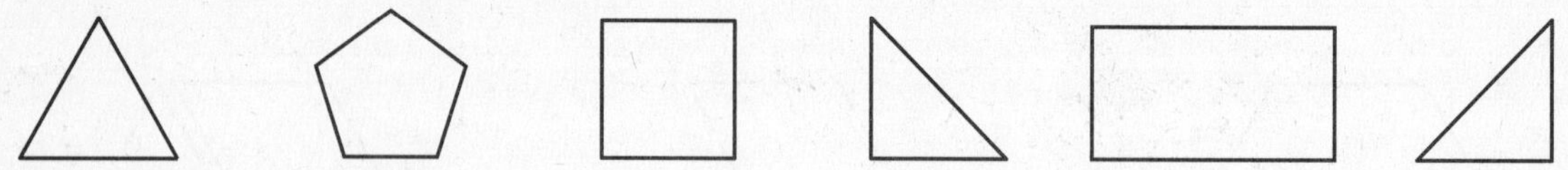

5. 5 or more sides

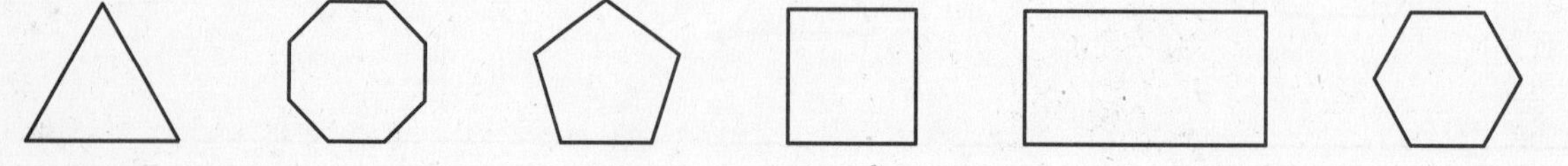

Problem Solving

Solve.

6. I have 6 flat faces. I have 12 sides. What am I?

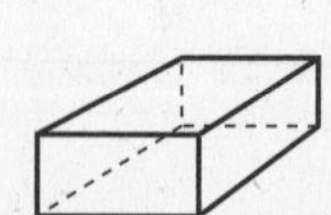
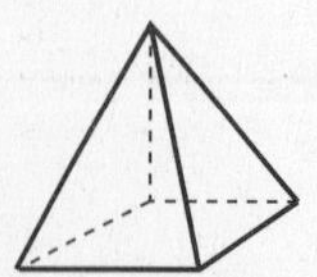

7. I have no sides and no vertices. What am I?

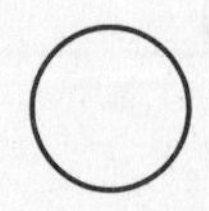

Name ____

Same Size and Same Shape

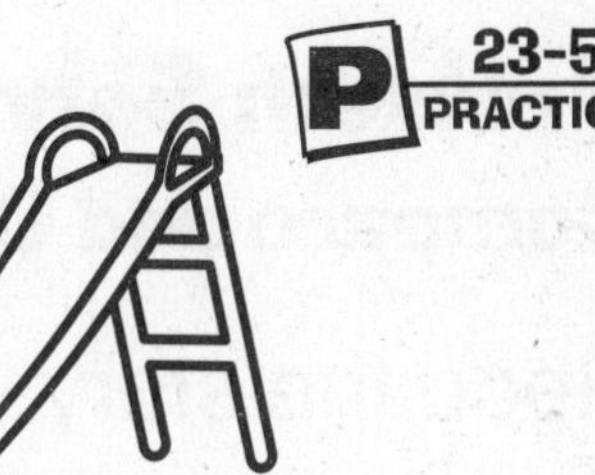

Color the shapes that match in each row. Write the letters under the figures you colored in order at the bottom.

Why did the chicken cross the playground?

1. c t a h k h

2. e z o d y t

3. h t f e u r

4. p s w l w h

5. i t s d e h

To get to ___ ___ ___ ___ ___ ___ ___ ___

___ ___ ___ ___ ___.

Name ____________________

Problem Solving Skill: Reading for Math

Compare and Contrast

Read the story.

Roberto made a model of a playground. He used 3-dimensional figures for some things in his model. He also used 2-dimensional shapes in his model.

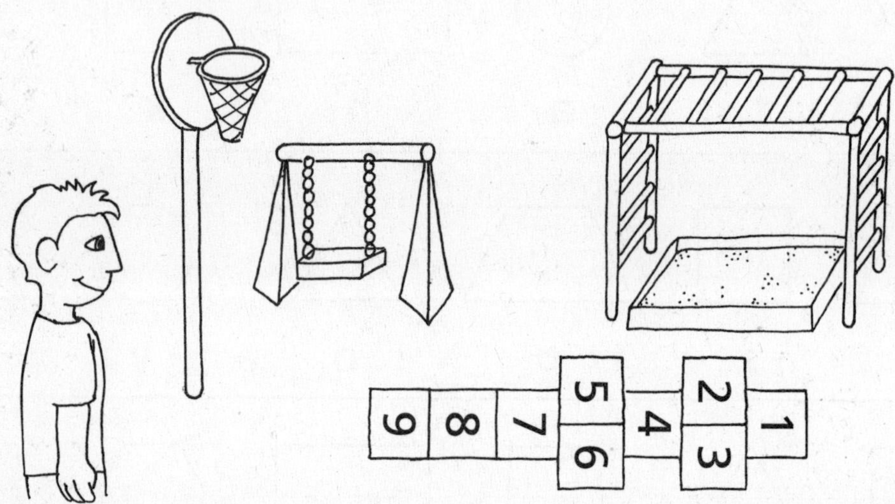

1. Which shape did Roberto use in drawing the hopscotch outline?

   triangle   square   rectangle

2. Name two objects in the model that are made from cylinders.

   ____

   ____

3. How are the two objects the same? ____________________

   ____

4. How are the two objects different?

   ____

   ____

Name ______________________________

Position

Position words tell where things are.

Draw.

1.  going up 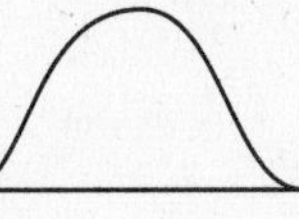

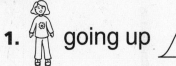

2. 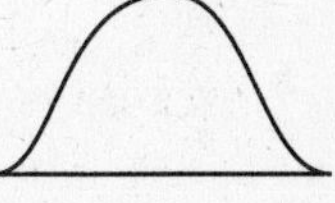 going down

3.  below 

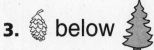

4.  near 

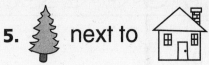

5.  next to

6.  above 

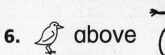

Name ____________________

Open and Closed Shapes

1. Color all the closed shapes.

2. Draw 3 open shapes and 3 closed shapes in the picture frame.

3. Color the closed figures in your picture.

Name ____________________

Reflections of a Shape

You may use  . Draw each reflection.

1. 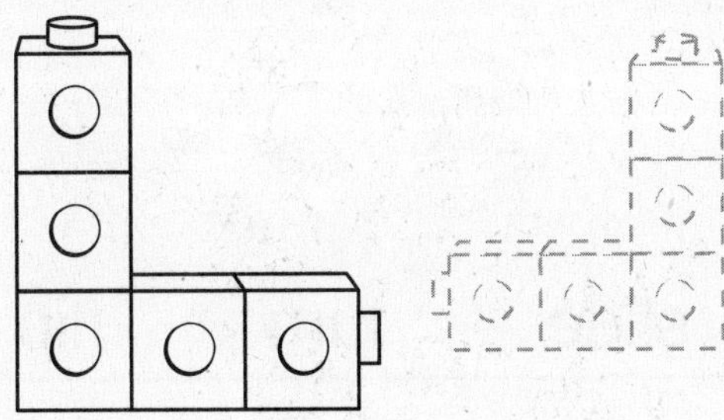

2. 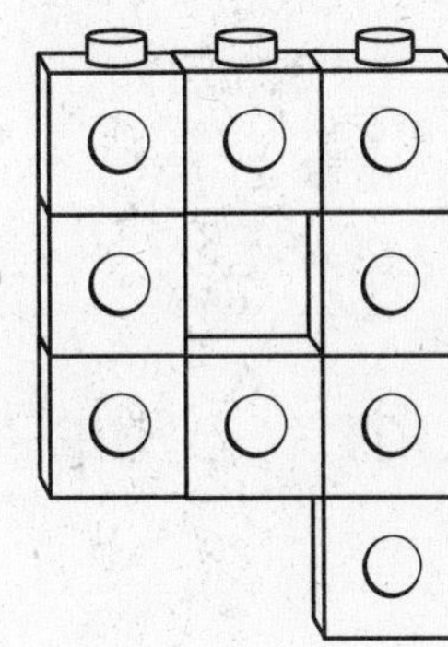

3. 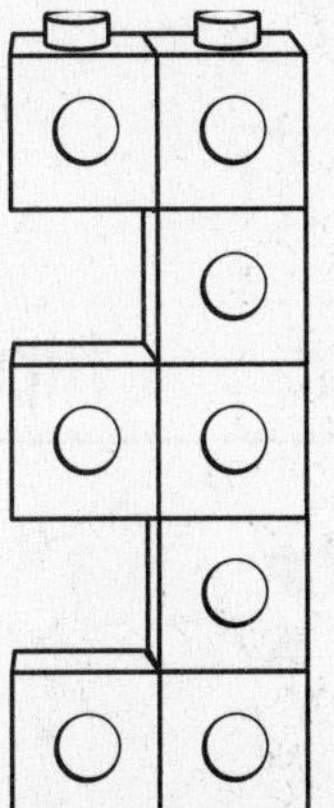

4. 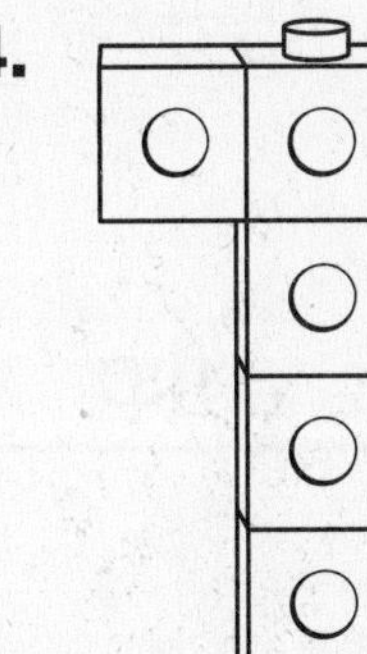

5. 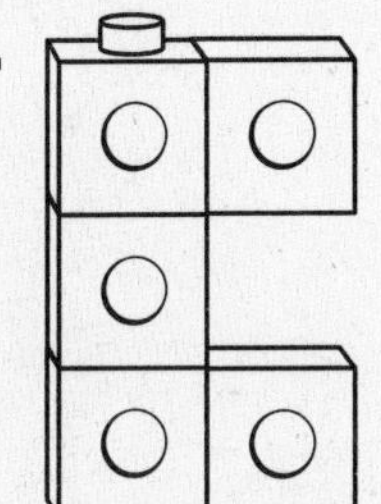

6. 

Name ____

Slides, Turns, and Flips

P 24-4 PRACTICE

Move  as shown.

Circle the word that tells the move you made.

1. 

slide turn flip

2.   

slide turn flip

3. 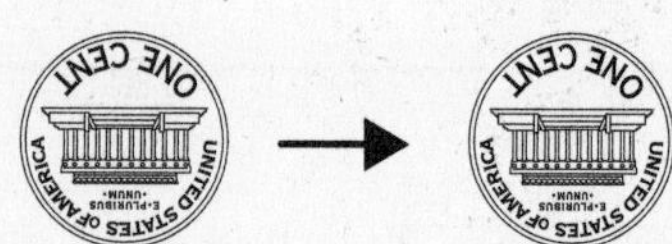

slide turn flip

4. 

slide turn flip

5.  

slide turn flip

6.  

slide turn flip

Problem Solving

Solve.

Circle the  that shows the move.

7. 

slide

8. 

turn

Name ____________________

Symmetry

Color the shapes that have symmetry.

1. 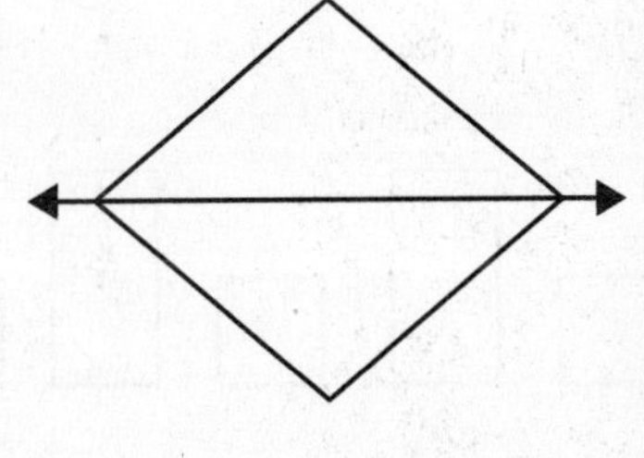

2. 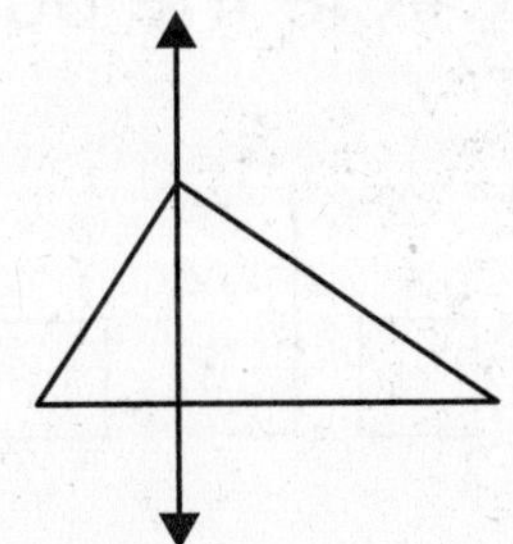

3. 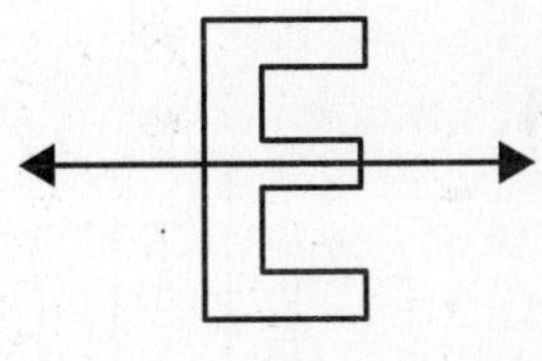

4. 

5. 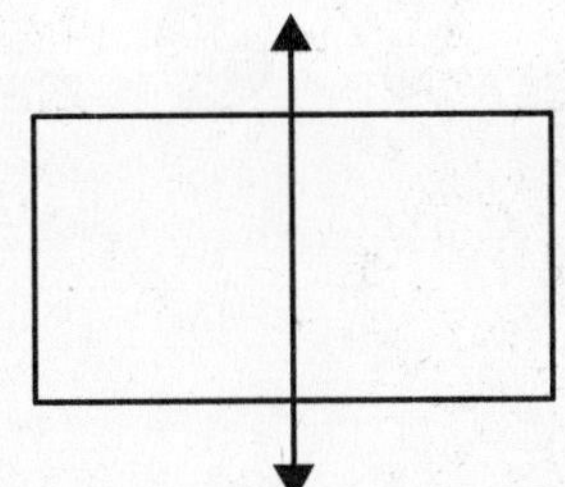

6. 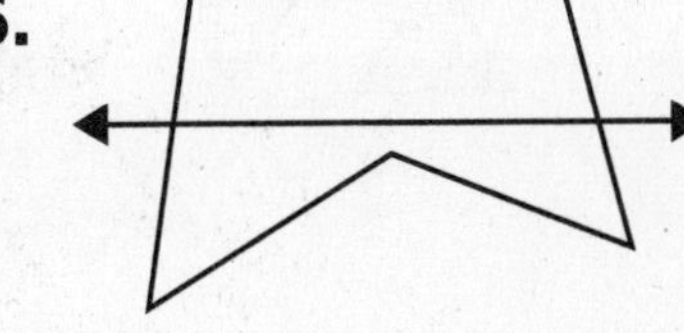

7. 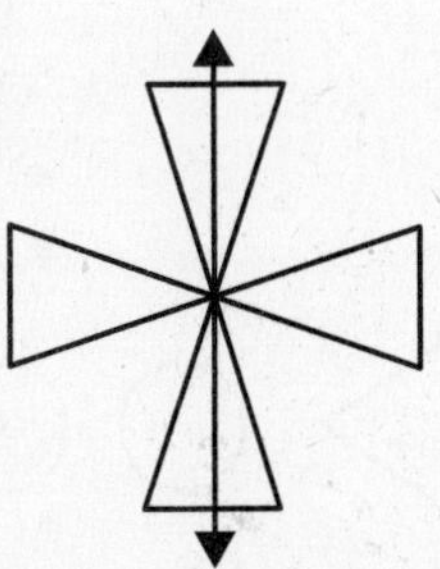

8. 

9. 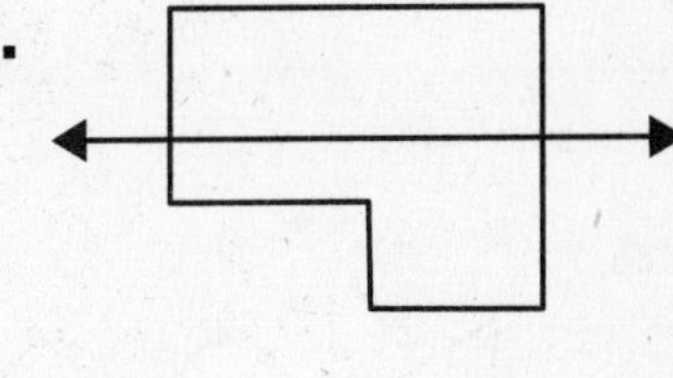

10. 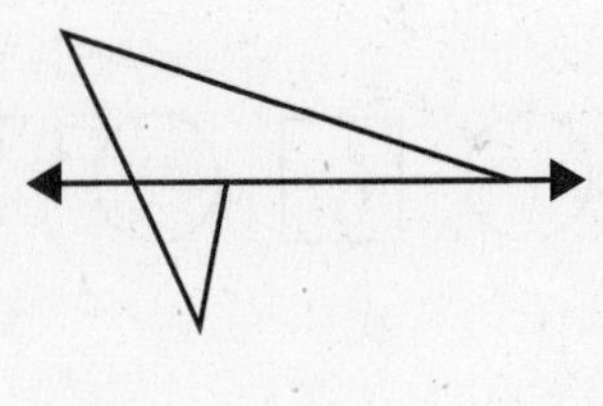

11. 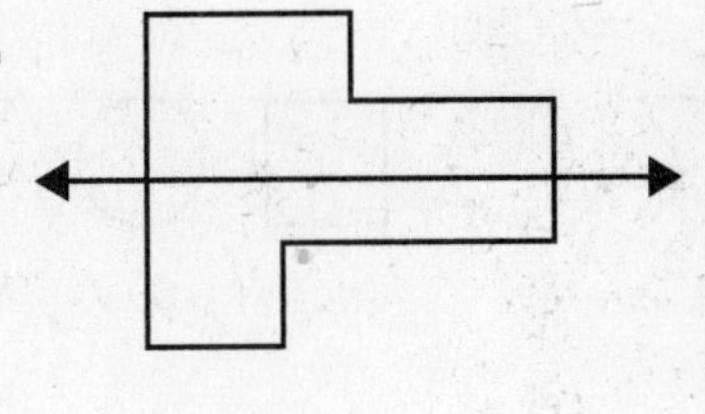

12. 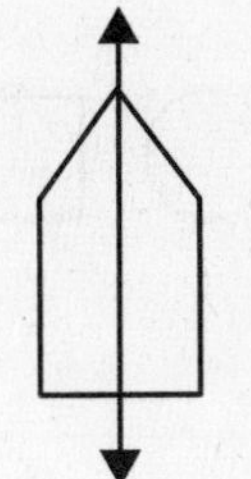

Name ______________________________

Problem Solving: Strategy

Describe Patterns • Algebra

Look at each pattern. Circle the pattern unit.

1.

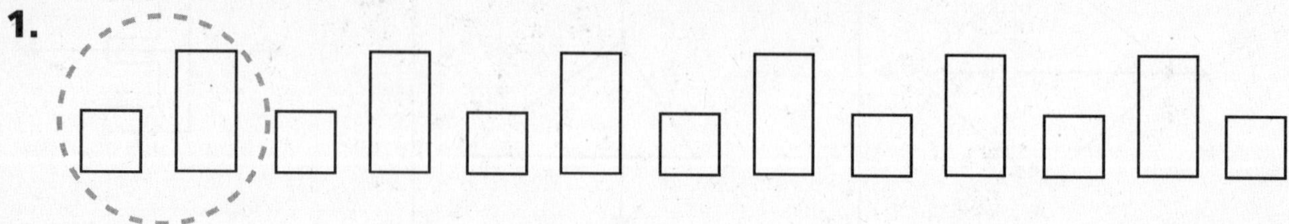

2.

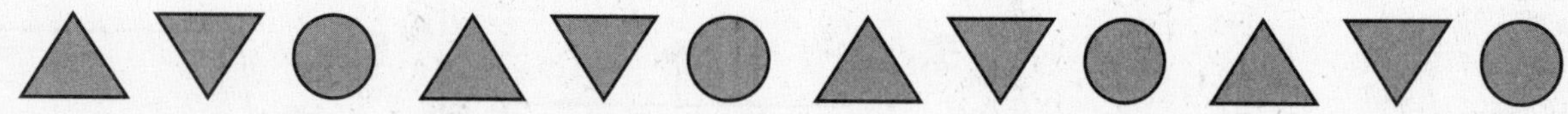

3.

4.

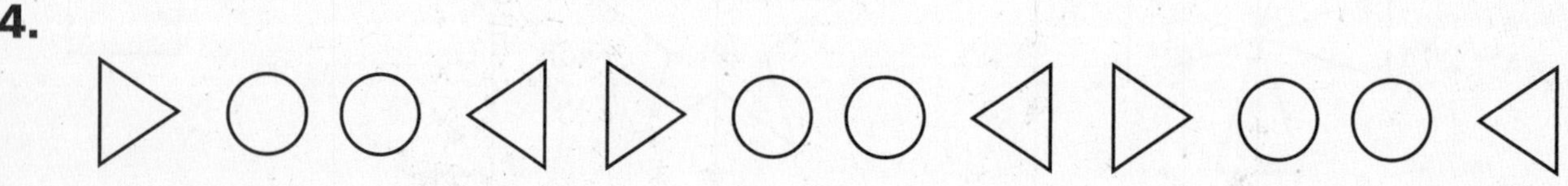

5.

Use with Grade 1, Chapter 24, Lesson 6, pages 439–440.

Name ______________________________

Explore Fractions

Circle each picture that shows equal parts.

1. 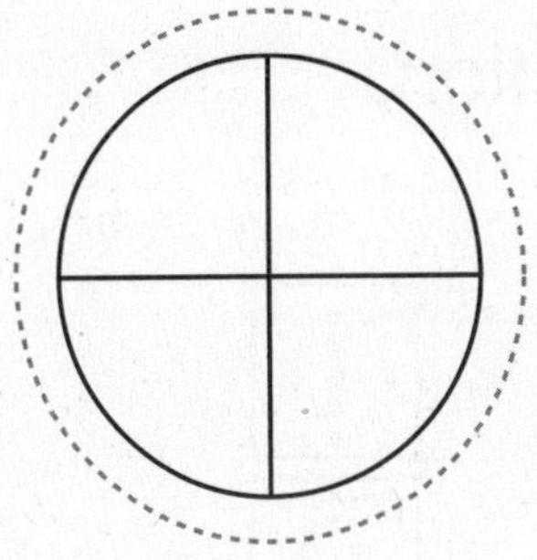

2. 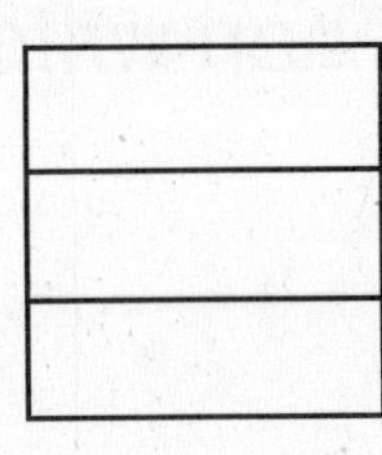

3. 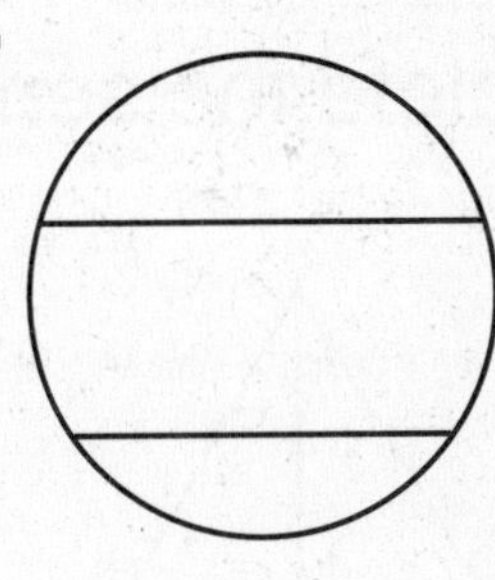

4. 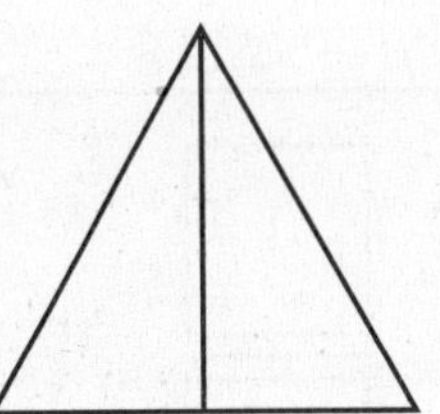

5. 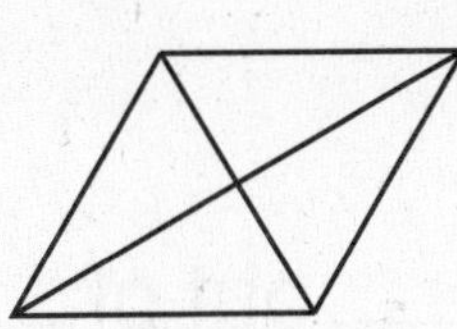

6. 

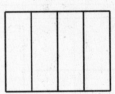

7. 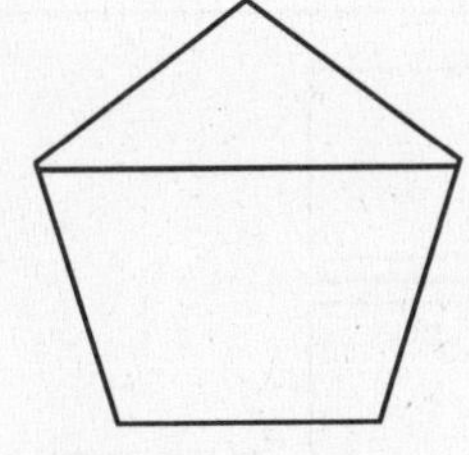

8. 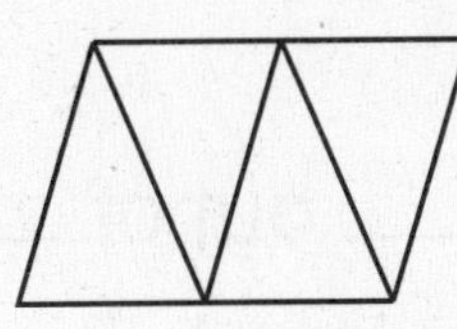

9. 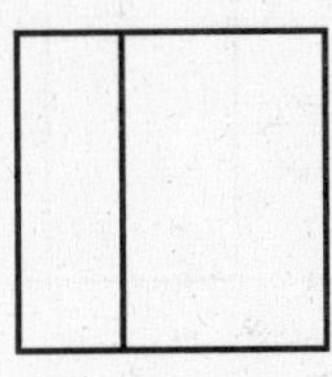

Problem Solving

Solve. Draw a picture.

10. 4 friends want to share the sandwich. Each friend wants an equal part. Draw lines to show where you would cut the sandwich.

Name ____________________

Unit Fractions

Complete.

Use green. Color one part.
Complete the sentence. Then write the fraction.

1. 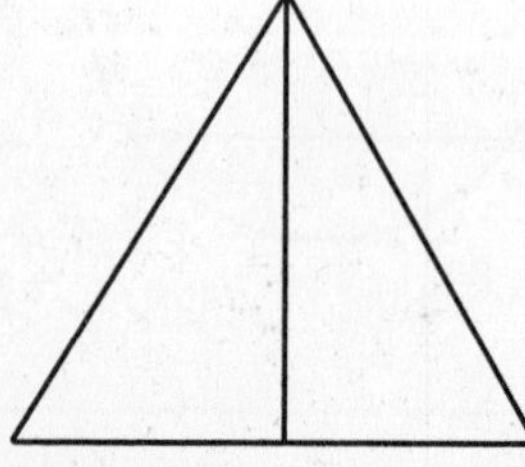

1 out of 2 equal parts is green

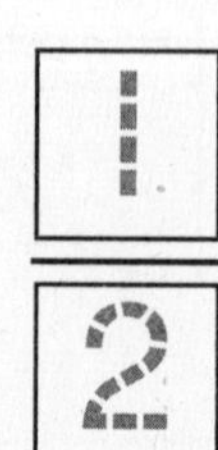

green

---

2. 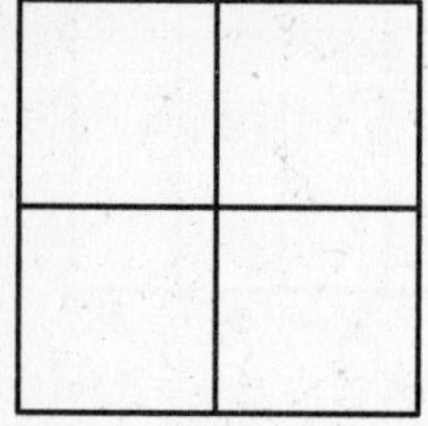

_____ out of _____ equal parts is green

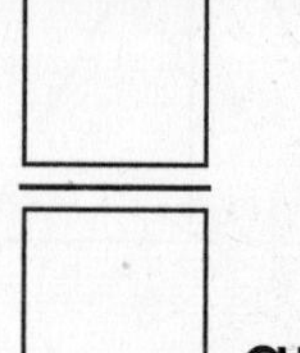
green

---

3. 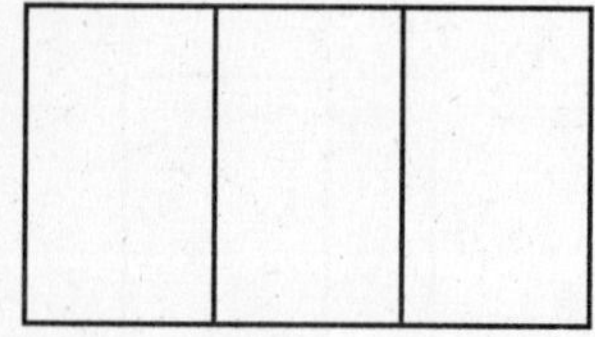

_____ out of _____ equal parts is green

green

---

4. 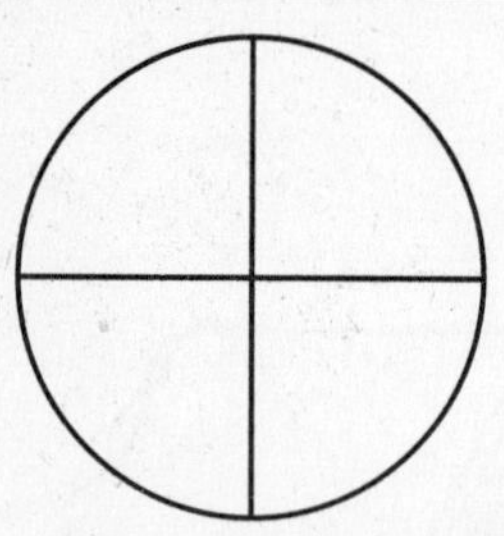

_____ out of _____ equal parts is green

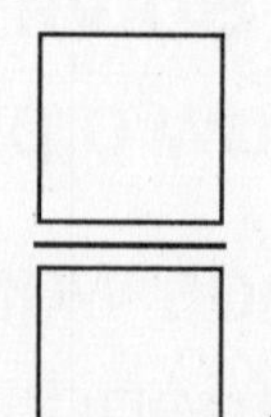
green

Use with Grade 1, Chapter 25, Lesson 2, pages 455–456.

Name ____

Other Fractions

Circle the correct fraction.

1. 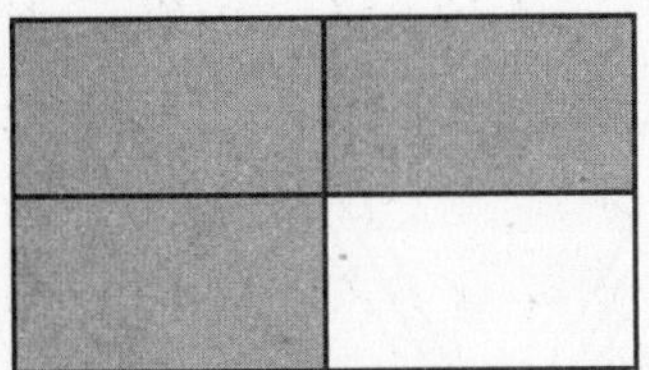

$\frac{1}{4}$ $\frac{2}{4}$ $\frac{3}{4}$

2. 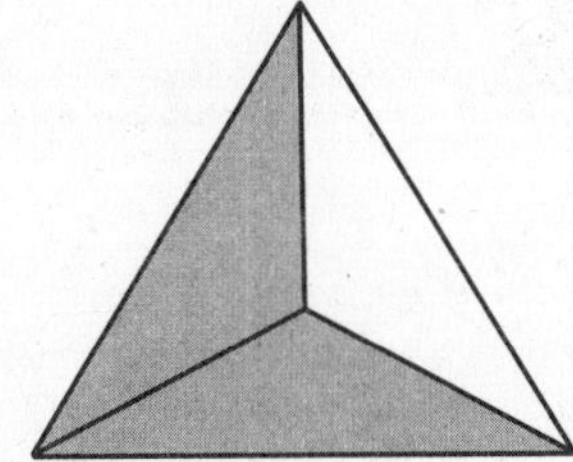

$\frac{1}{3}$ $\frac{2}{3}$ $\frac{3}{3}$

3. 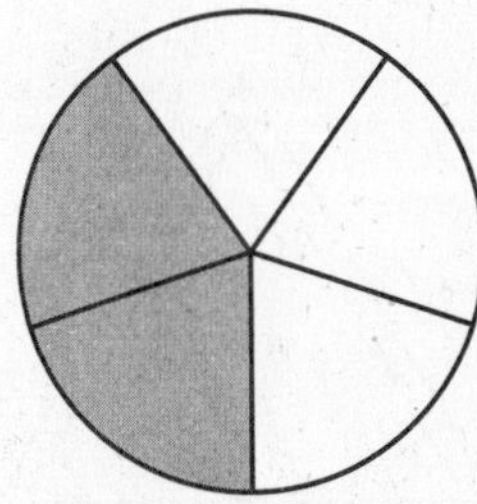

$\frac{2}{5}$ $\frac{3}{5}$ $\frac{4}{5}$

4. 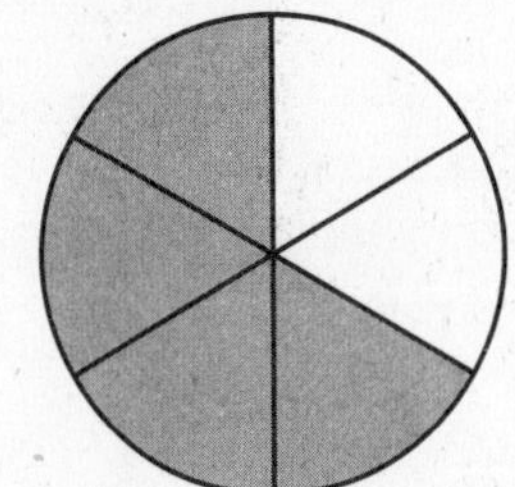

$\frac{4}{6}$ $\frac{5}{6}$ $\frac{6}{6}$

5. 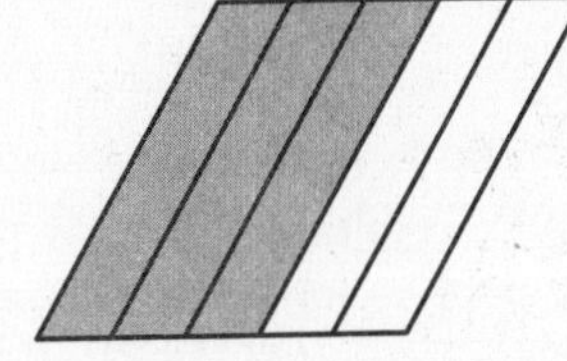

$\frac{2}{5}$ $\frac{3}{5}$ $\frac{4}{5}$

6. 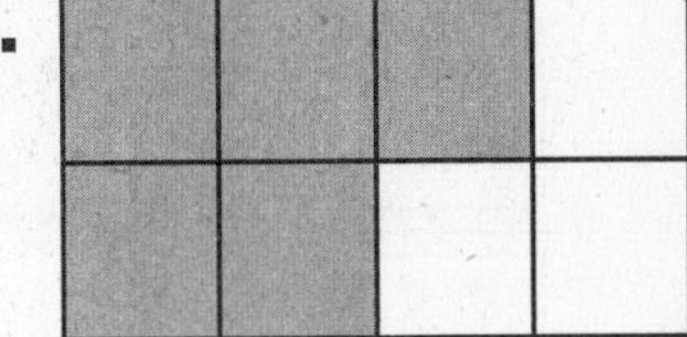

$\frac{2}{8}$ $\frac{5}{8}$ $\frac{7}{8}$

7. 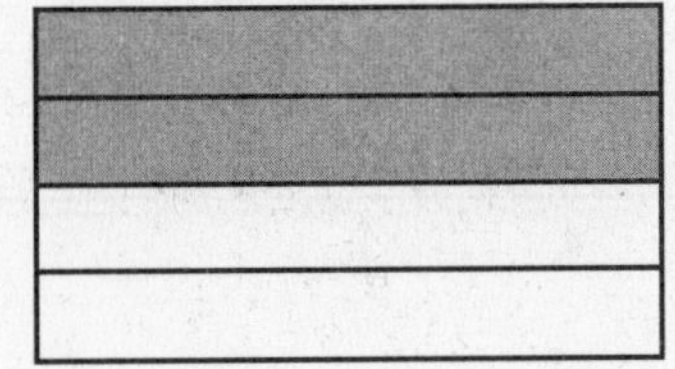

$\frac{2}{4}$ $\frac{3}{4}$ $\frac{4}{4}$

8. 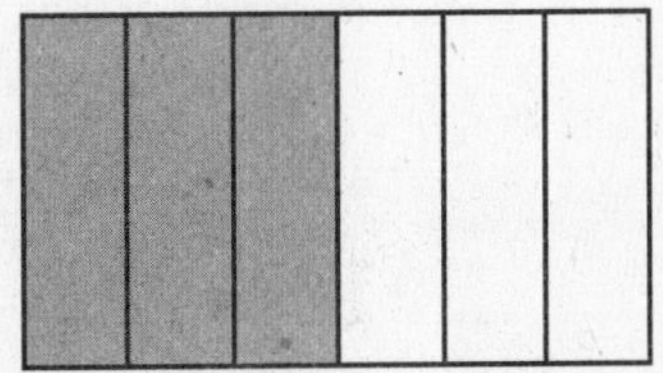

$\frac{2}{6}$ $\frac{3}{6}$ $\frac{5}{6}$

9. 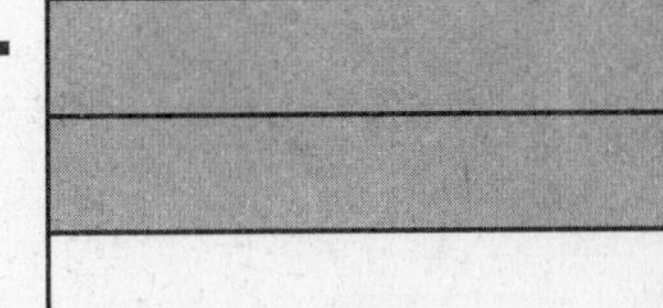

$\frac{1}{3}$ $\frac{2}{3}$ $\frac{3}{3}$

Name ____________________

Fractions Equal to 1

P 25-4 PRACTICE

You may use pattern blocks. Write the fraction for the whole.

1.

$\frac{2}{2}$ = 1 whole

2.

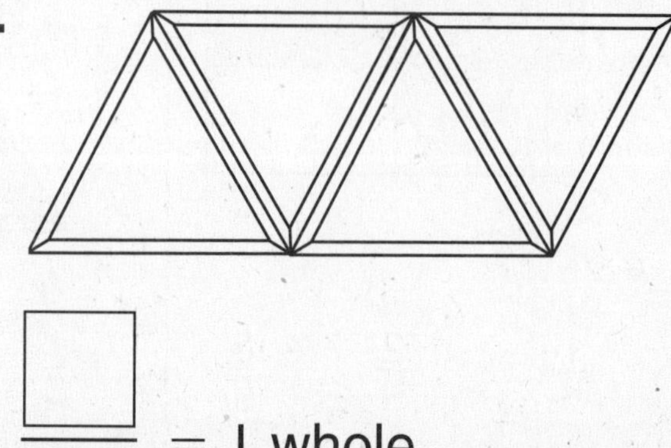

$\frac{\square}{\square}$ = 1 whole

3.

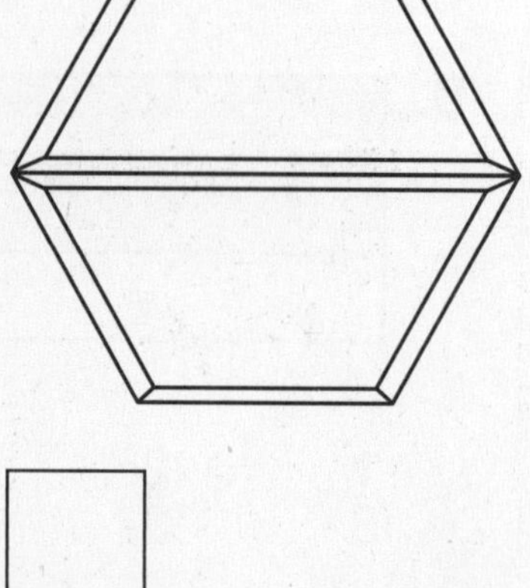

$\frac{\square}{\square}$ = 1 whole

4.

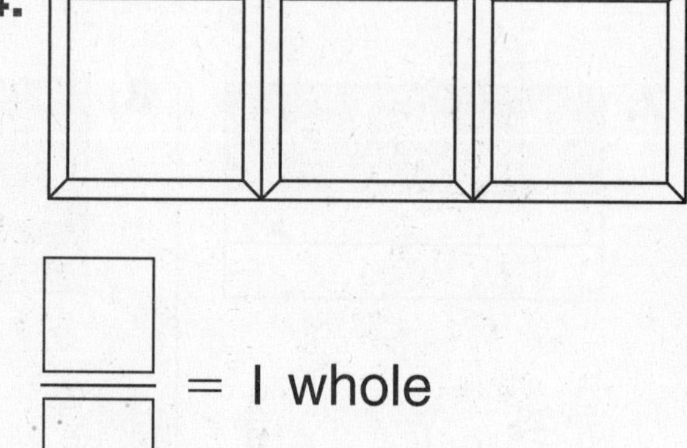

$\frac{\square}{\square}$ = 1 whole

Name ________________________________

Fractions of a Group

Write how many are in the group. Circle 1 part of the group. Write the fraction that names the part you circled.

1. 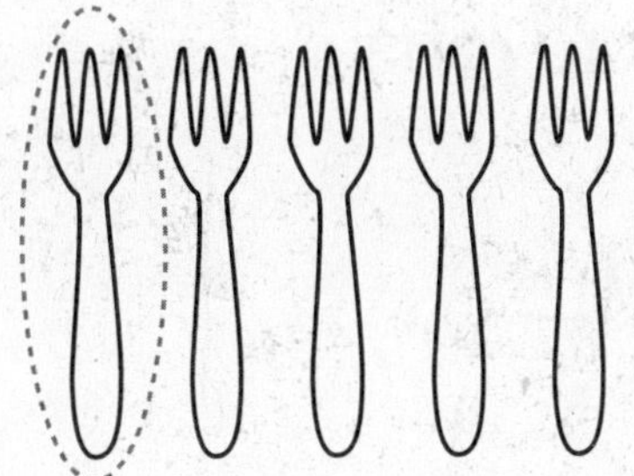

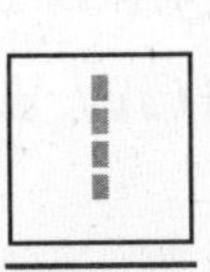

5 forks

2. 

_____ pies

3. 

_____ apples

4. 

_____ students

5. 

_____ flowers

6. 

_____ umbrellas

Name ______________________________

Problem Solving Skill: Reading for Math

Use Illustrations

Read the story.

Ryan, Jane, and Joe are making pictures. Ryan and Jane like to paint. Joe likes to draw with crayons.

Use the illustration to solve.

1. Joe's paper is divided into ________.
2. How much of the paper has Joe used? ______
3. How many students are standing? Write the fraction. ______
4. How much of the paper has Jane used? Write the fraction. ______

Name ______________________________

Compare Fractions

26-1 PRACTICE

Color the food to show each fraction.
Circle the greater fraction.

Compare. Which one is larger?

1. $\frac{3}{4}$

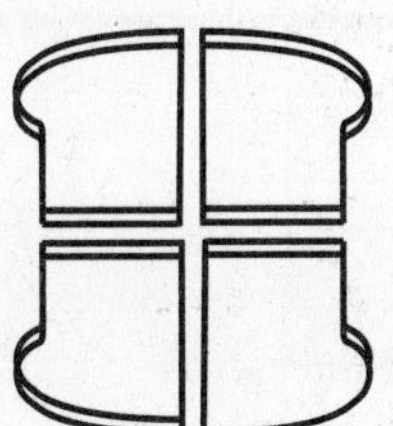

$\frac{2}{4}$

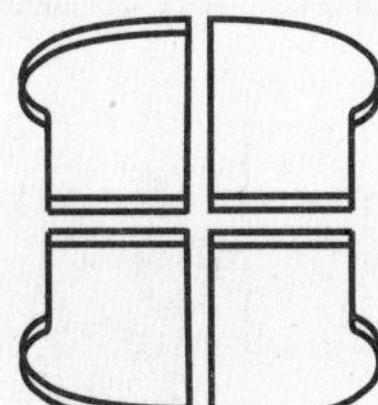

2. $\frac{2}{6}$

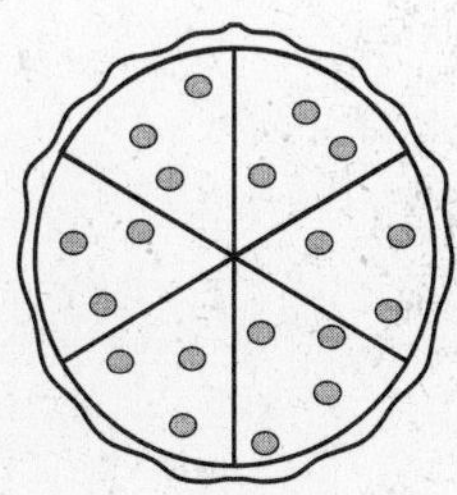

$\frac{5}{6}$

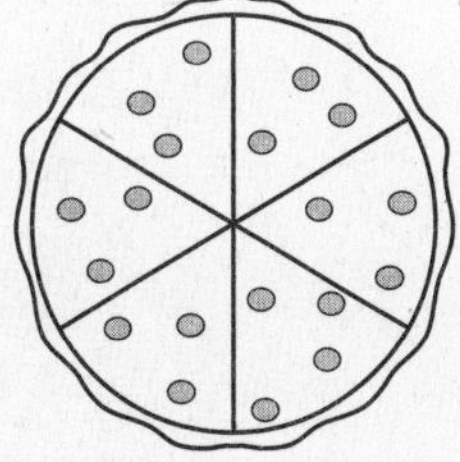

3. $\frac{1}{8}$

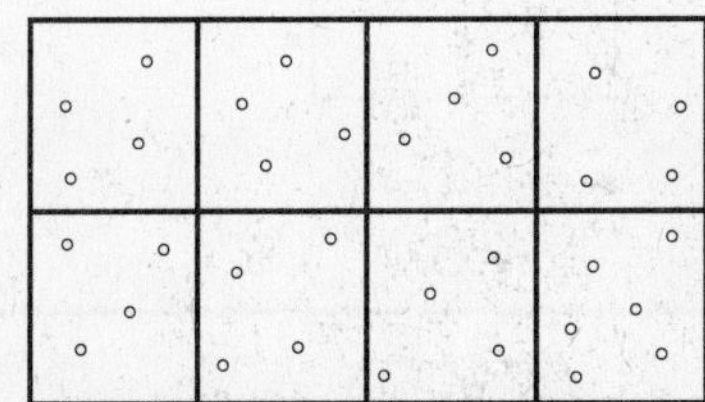

$\frac{3}{8}$

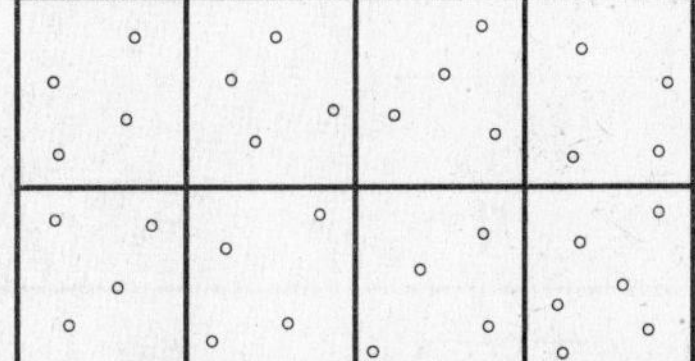

4. $\frac{4}{5}$

$\frac{3}{5}$

Name ______________________________

Compare Unit Fractions

P
26-2
PRACTICE

Color 1 part to show each fraction.
Circle the fraction that names the larger part.

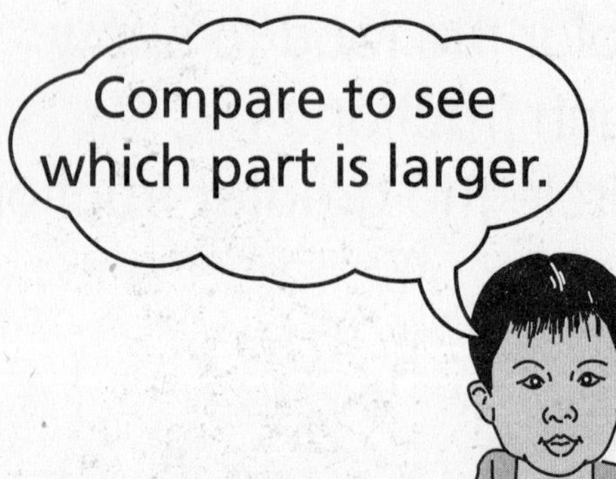
Compare to see which part is larger.

1.

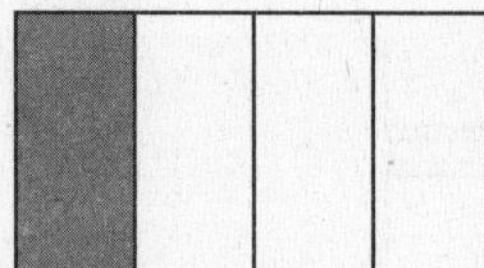

$\frac{1}{4}$

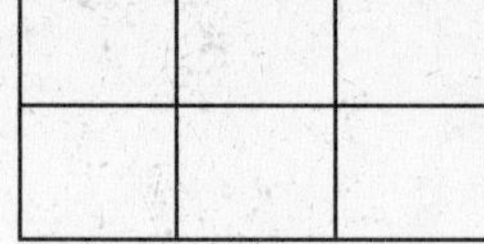

$\frac{1}{6}$

2.

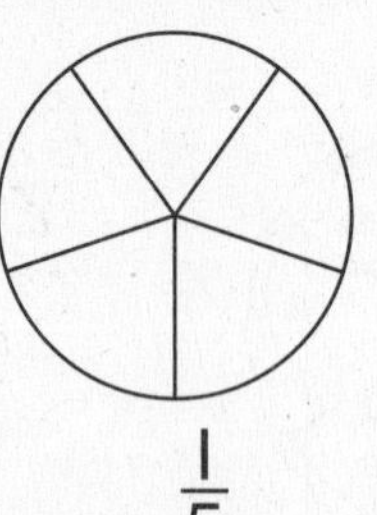

$\frac{1}{5}$

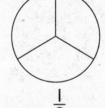

$\frac{1}{3}$

3.

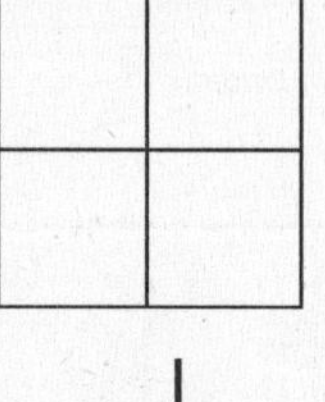

$\frac{1}{4}$

$\frac{1}{8}$

4.

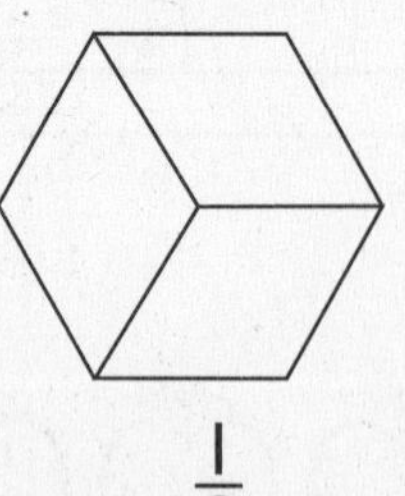

$\frac{1}{3}$

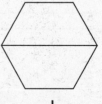

$\frac{1}{2}$

Name ____________________

Equally Likely

Some spinners are equally likely to land on either light or dark. Circle them.

1.

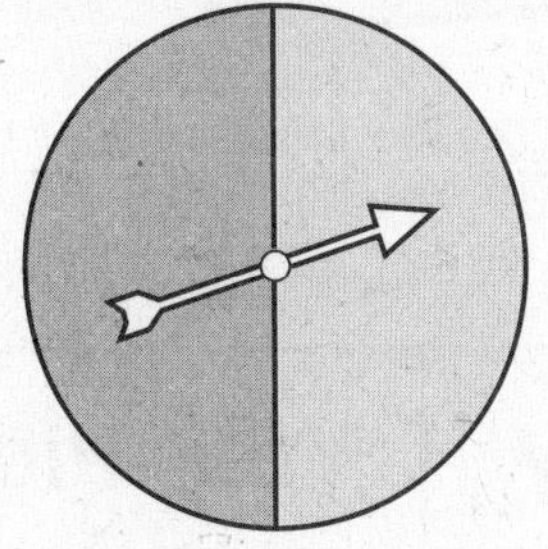

2.

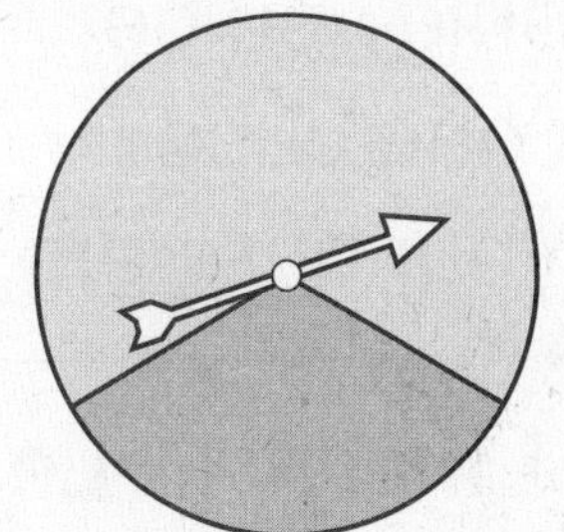

3.

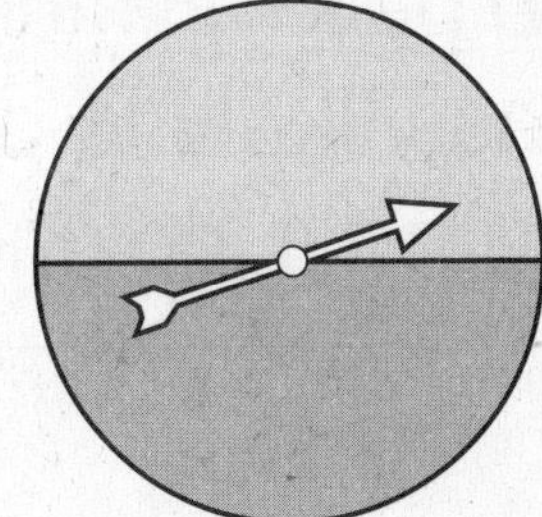

4.

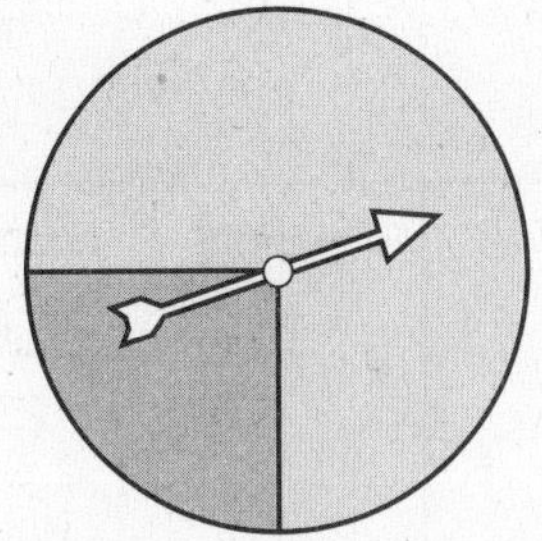

5.

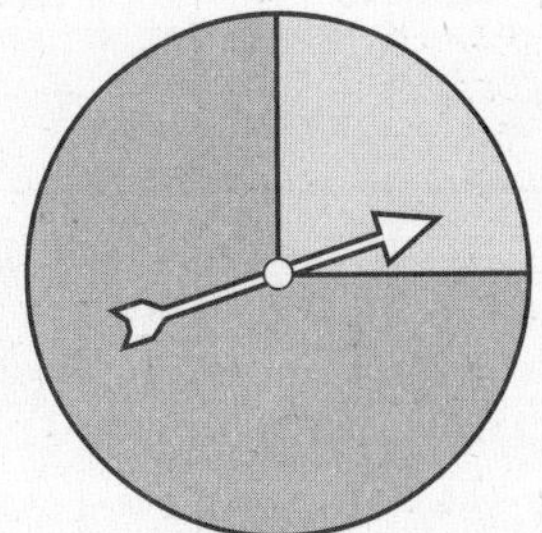

6.

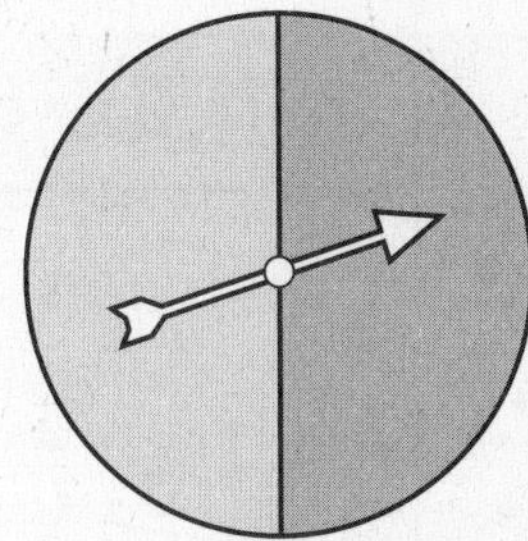

7.

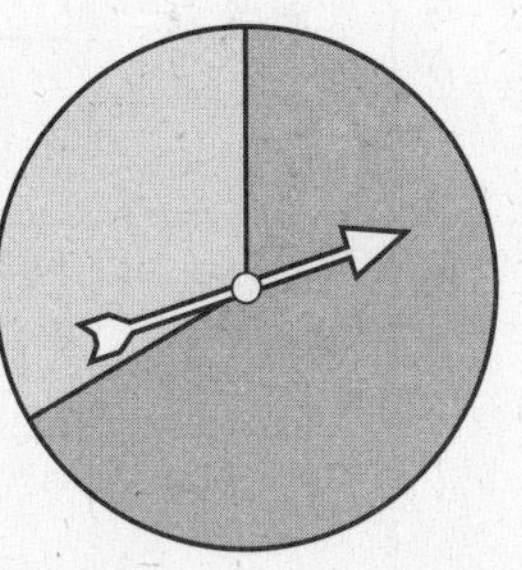

8.

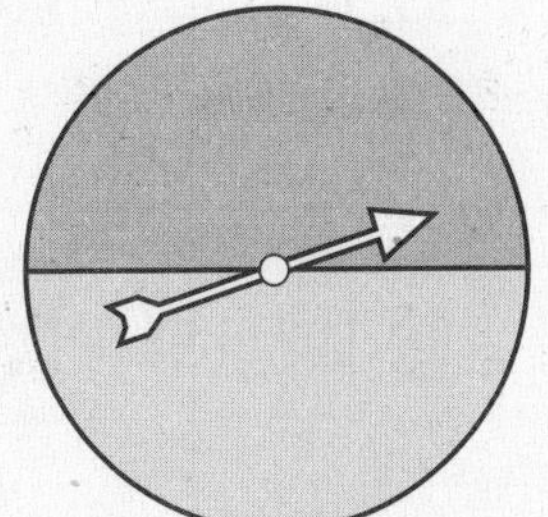

9.

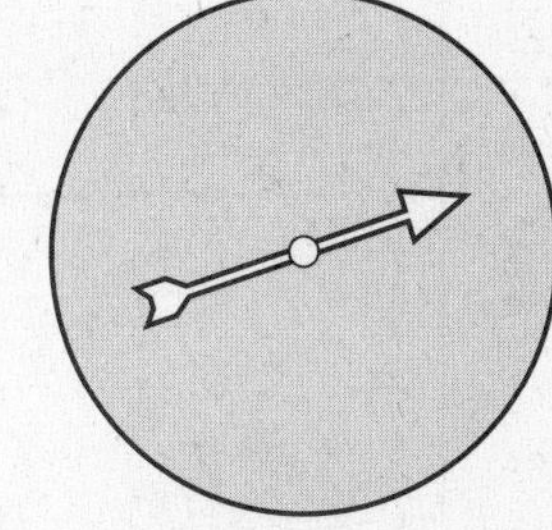

10.

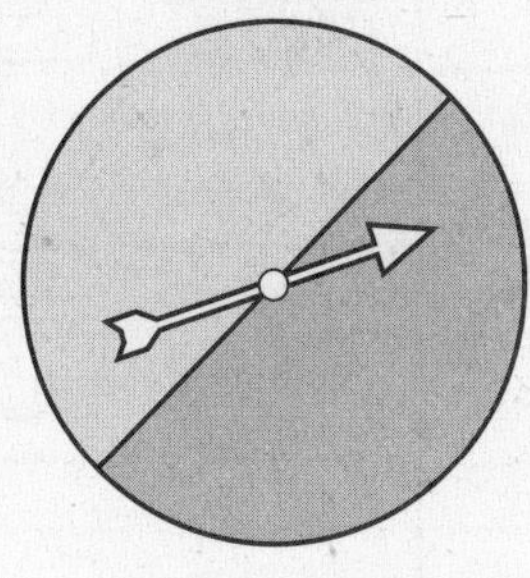

11.

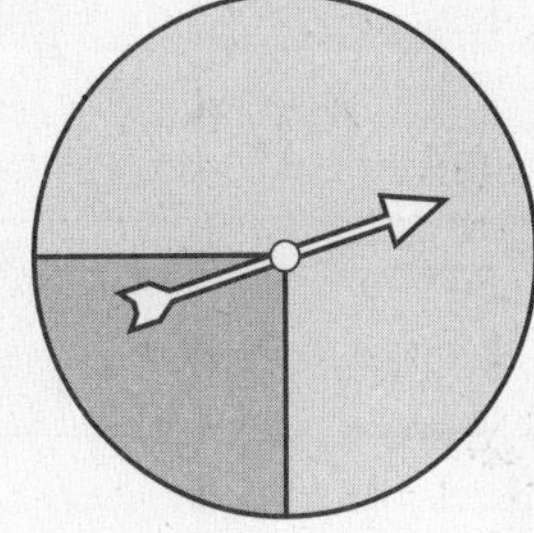

12.

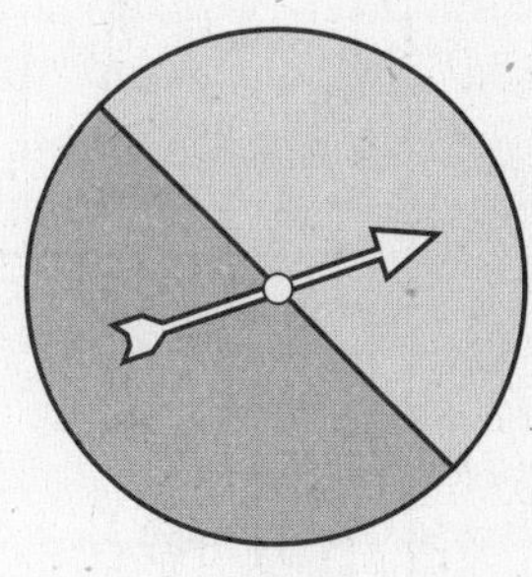

Name ____

More Likely and Less Likely

26-4 PRACTICE

Put cubes in a paper bag as shown. Which cube would you be more likely to pick? Which cube would you be less likely to pick? Pick one cube without looking. Circle your answers.

	Bag	More Likely	Less Likely	Your Pick
1.				
2.				
3.				
4.				
5.				
6.				

Use with Grade 1, Chapter 26, Lesson 4, pages 479–480.

Name ____________________

Certain, Probable, Impossible

Use (spinner). Color the (spinner) so that it is certain, probable, or impossible that the (arrow) will land on red.

1. certain

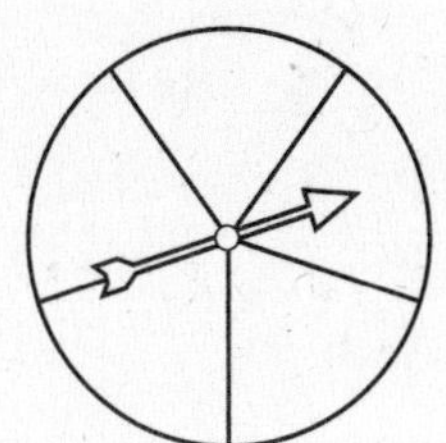

2. probable

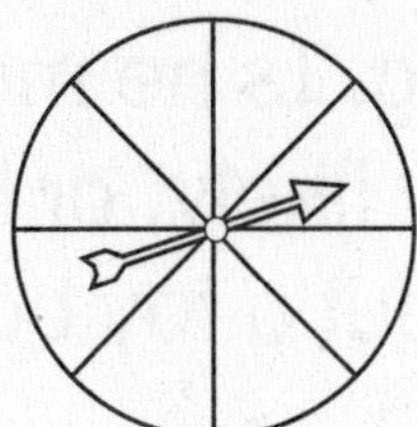

3. probable

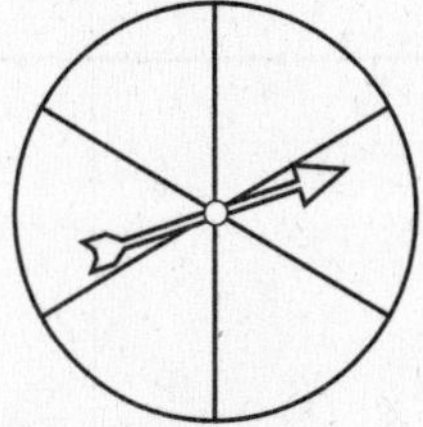

4. impossible

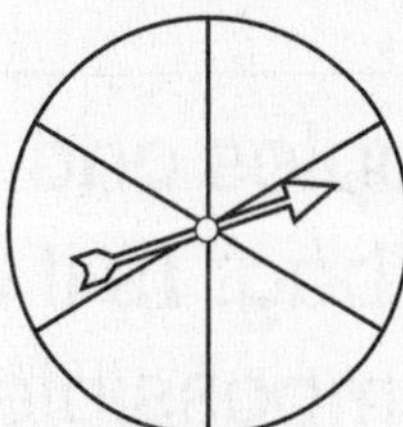

5. certain

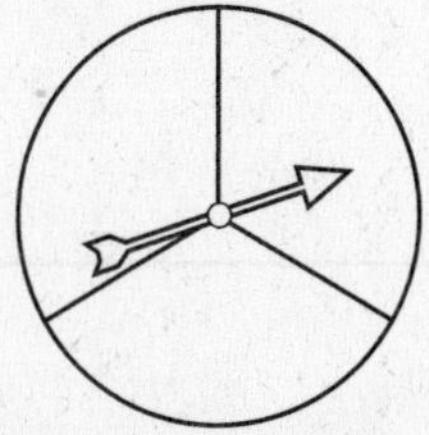

6. probable

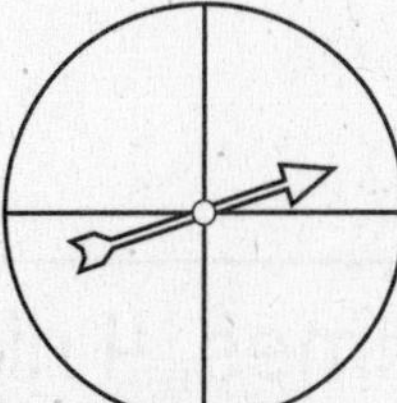

7. impossible

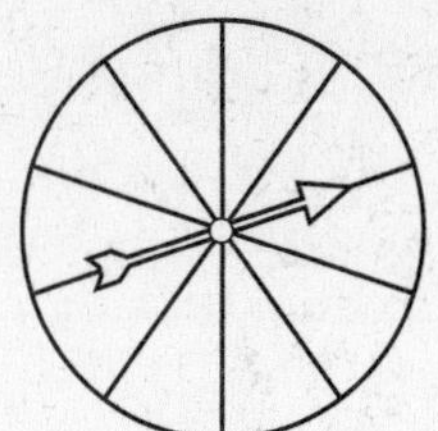

8. probable

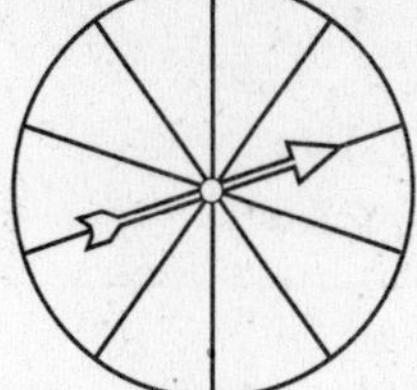

Name ______________________________

Problem Solving: Strategy

Draw a Picture

26-6 PRACTICE

Draw a picture to solve.

Circle your answer. | Draw or write to explain.

1. Tony has 5 toy cars and 3 toy boats in a bag. Is he more likely, equally likely, or less likely to pull out a toy boat?

certain
equally likely
less likely

boat

2. Jess has 3 apples and 4 bananas in a bag. Is it certain, probable, or impossible that she will pull out a banana.

certain
probable
impossible

banana

3. Bode has 3 dimes, 4 nickels and 1 quarter in a bag. Is it certain, probable, or impossible that he will pull out a penny.

certain
probable
impossible

penny

Name ______________________________

Patterns with 10

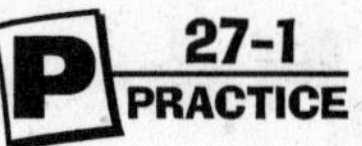

Add. Use ○ and [ten frame].

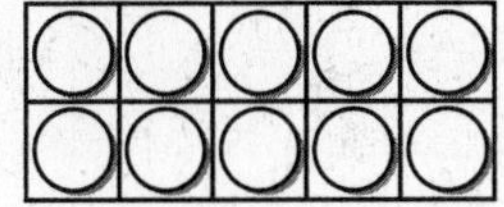 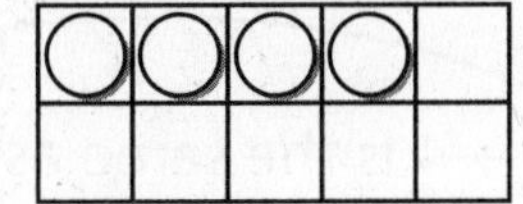

1. 10 + 4 = 14

2. 10 + 6 = ____ 3. 10 + 7 = ____ 4. 10 + 2 = ____

5. 10 + 1 = ____ 6. 10 + 9 = ____ 7. 10 + 5 = ____

8. $10 + 3$ 9. $10 + 8$ 10. $10 + 0$ 11. $10 + 6$

12. $2 + 10$ 13. $5 + 10$ 14. $9 + 10$ 15. $10 + 10$

Problem Solving

Solve.

16. Maria puts 18 ○ in two boxes. She puts 10 ○ in the first box. How many ○ does she put in the second box?

____

17. Will puts 16 (penny) in two rows. There are 6 (penny) in the second row. How many (penny) are in the first row?

____

Name ____________________

Make a 10 to Add

Add. Use a ◯, ●, and ▭.

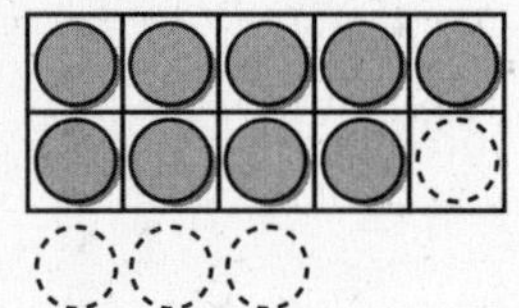

9 + 4 is the same as 9 + 1 + 3.
9 + 4 is the same as 10 + 3.
9 + 4 and 10 + 3 equal 13.

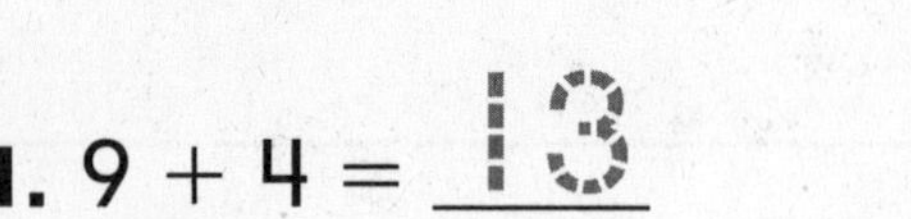

1. 9 + 4 = 13

2. 7 + 5 = _____ 3. 9 + 7 = _____ 4. 8 + 4 = _____

5. 6 + 7 = _____ 6. 8 + 6 = _____ 7. 7 + 8 = _____

8. 7 + 4 = _____ 9. 9 + 6 = _____ 10. 8 + 8 = _____ 11. 9 + 5 = _____ 12. 7 + 6 = _____ 13. 6 + 9 = _____

Problem Solving

Solve. Use ◯ and ▭.

14. 9 + 8 is the same as 10 + _____.

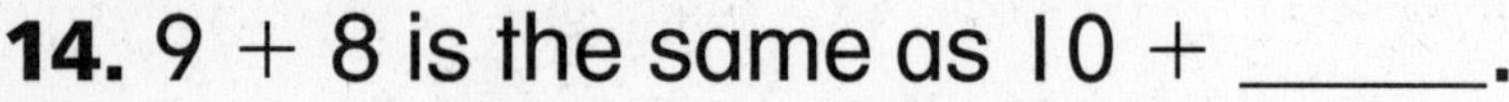

15. 7 + 7 is the same as 10 + _____.

16. 8 + 5 is the same as 10 + _____.

17. 9 + 9 is the same as 10 + _____.

Name ______________________________

Relate Addition and Subtraction • Algebra

Use [cube] and [cube]. Add. Then write the related subtraction facts. Subtract.

Add	Subtract
7 + 5 = 12	12 − 5 = 7
	12 − 7 = 5

1. 9 + 4 = ____
   ____ ◯ ____ = ____
   ____ ◯ ____ = ____

2. 6 + 5 = ____
   ____ ◯ ____ = ____
   ____ ◯ ____ = ____

3. 8 + 7 = ____
   ____ ◯ ____ = ____
   ____ ◯ ____ = ____

4. 4 + 8 = ____
   ____ ◯ ____ = ____
   ____ ◯ ____ = ____

5. 3 + 9 = ____
   ____ ◯ ____ = ____
   ____ ◯ ____ = ____

6. 9 + 6 = ____
   ____ ◯ ____ = ____
   ____ ◯ ____ = ____

Name ______________________________

Use Addition to Subtract • Algebra

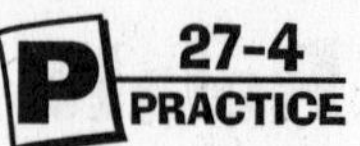

Find each missing number.

1. 15 − 8 = __7__

__7__ + 8 = 15

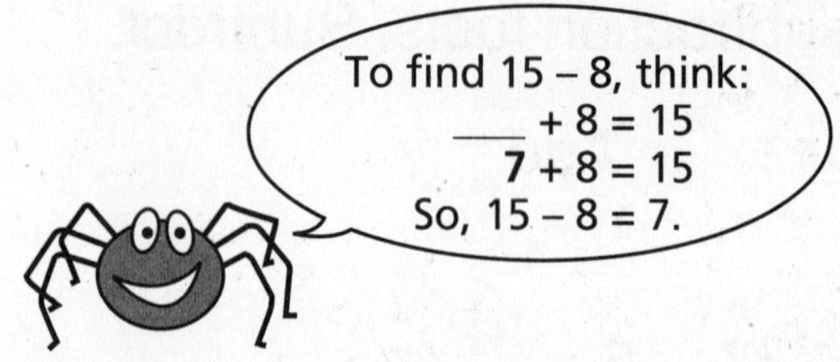

To find 15 – 8, think:
___ + 8 = 15
7 + 8 = 15
So, 15 – 8 = 7.

2. 13 − 4 = ☐

☐ + 4 = 13

3. 12 − 5 = ☐

☐ + 5 = 12

4. 14 − 6 = ☐

☐ + 6 = 14

5. 11 − 8 = ☐

☐ + 8 = 11

6. 18 − 9 = ☐ ; ☐ + 9 = 18

7. 15 − 8 = ☐ ; ☐ + 8 = 15

8. 12 − 3 = ☐ ; ☐ + 3 = 12

9. 13 − 5 = ☐ ; ☐ + 5 = 13

10. 14 − 7 = ☐ ; ☐ + 7 = 14

11. 12 − 8 = ☐ ; ☐ + 8 = 12

Use with Grade 1, Chapter 27, Lesson 4, pages 497–498.

Name ______________________________

Fact Families

Add or subtract.
Complete each fact family.

1. 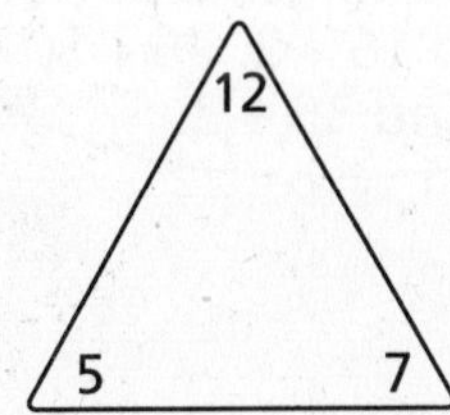

5 + 7 = 12   12 − 7 = 5

7 + 5 = 12   12 − 5 = 7

2. 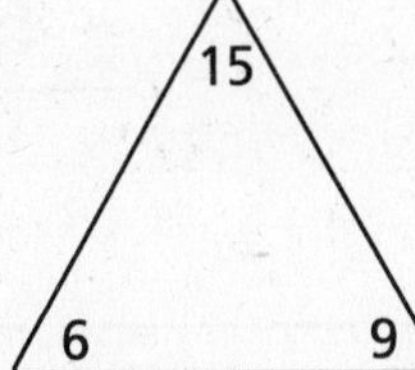

6 + 9 = ____   15 − 9 = ____

____ + ____ = ____   ____ − ____ = ____

3. 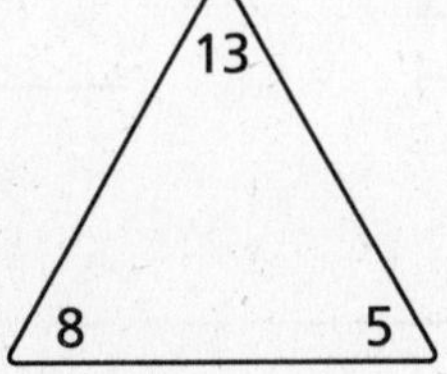

8 + 5 = ____   ____ − ____ = ____

____ + ____ = ____   ____ − ____ = ____

4. 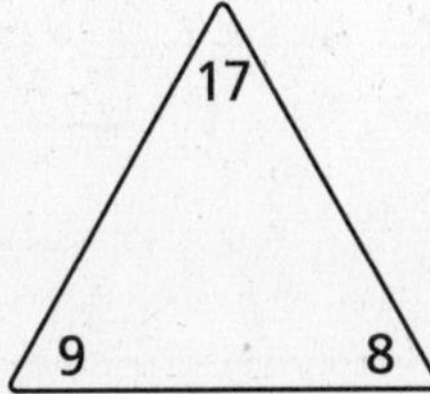

____ + ____ = ____   ____ − ____ = ____

____ + ____ = ____   ____ − ____ = ____

5. 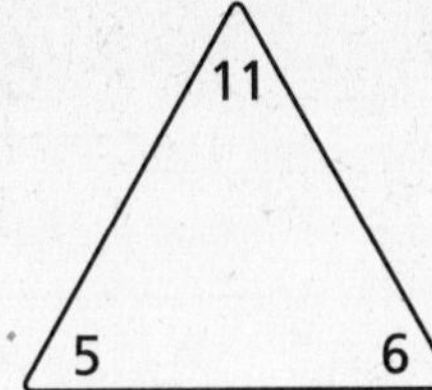

____ + ____ = ____   ____ − ____ = ____

____ + ____ = ____   ____ − ____ = ____

Name ________________

Addition and Subtraction Patterns • Algebra

Use a pattern to add or subtract.

1.
$$\begin{array}{r} 6 \\ +3 \\ \hline 9 \end{array} \quad \begin{array}{r} 16 \\ +3 \\ \hline 19 \end{array} \quad \begin{array}{r} 26 \\ +3 \\ \hline 29 \end{array} \quad \begin{array}{r} 36 \\ +3 \\ \hline 39 \end{array}$$

If you know 6 + 3 = 9, then you can use a pattern to find 16 + 3, 26 + 3, and 36 + 3.

2.
$$\begin{array}{r} 9 \\ -5 \\ \hline \end{array} \quad \begin{array}{r} 19 \\ -5 \\ \hline \end{array} \quad \begin{array}{r} 29 \\ -5 \\ \hline \end{array} \quad \begin{array}{r} 39 \\ -5 \\ \hline \end{array} \quad \begin{array}{r} 49 \\ -5 \\ \hline \end{array} \quad \begin{array}{r} 59 \\ -5 \\ \hline \end{array}$$

3.
$$\begin{array}{r} 6 \\ +4 \\ \hline \end{array} \quad \begin{array}{r} 16 \\ +4 \\ \hline \end{array} \quad \begin{array}{r} 26 \\ +4 \\ \hline \end{array} \quad \begin{array}{r} 36 \\ +4 \\ \hline \end{array} \quad \begin{array}{r} 46 \\ +4 \\ \hline \end{array} \quad \begin{array}{r} 56 \\ +4 \\ \hline \end{array}$$

4.
$$\begin{array}{r} 8 \\ -7 \\ \hline \end{array} \quad \begin{array}{r} 18 \\ -7 \\ \hline \end{array} \quad \begin{array}{r} 28 \\ -7 \\ \hline \end{array} \quad \begin{array}{r} 38 \\ -7 \\ \hline \end{array} \quad \begin{array}{r} 48 \\ -7 \\ \hline \end{array} \quad \begin{array}{r} 58 \\ -7 \\ \hline \end{array}$$

5.
$$\begin{array}{r} 9 \\ +6 \\ \hline \end{array} \quad \begin{array}{r} 19 \\ +6 \\ \hline \end{array} \quad \begin{array}{r} 29 \\ +6 \\ \hline \end{array} \quad \begin{array}{r} 39 \\ +6 \\ \hline \end{array} \quad \begin{array}{r} 49 \\ +6 \\ \hline \end{array} \quad \begin{array}{r} 59 \\ +6 \\ \hline \end{array}$$

Name ____________________

Problem Solving Skill: Reading for Math

Problem and Solution

Read the story.

There is a forest at the end of our street.

We made a path so we could walk in the forest.

The forest is full of plants and animals.

Solve.

1. What did they do so they could walk in the forest?

____

2. We saw 7 squirrels and 6 rabbits. How many animals did we see in all?

_____ animals

3. We saw 17 red birds and 6 blue birds. How many birds did we see in all?

_____ birds

4. We counted 27 big trees and 6 small trees. How many trees did we count in all?

_____ trees

Name ____________________

Add and Subtract Tens

Add or subtract. Use [tens rod].

1. 80 + 10 = ________

2. 30 + 30 = ________

3. 50 + 10 = ________

4. 60 + 20 = ________

5. 70 + 20 = ________

6. 40 + 30 = ________

7. 40 + 20 = ________

8. 50 − 30 = ________

9. 50 − 10 = ________

10. 70 − 20 = ________

11. 80 − 30 = ________

12. 30 − 20 = ________

13. 90 − 30 = ________

14. 60 − 10 = ________

Name ____________________

Count On Ones to Add

Count on to add.
Use the hundred chart.

1	2	3	4	5	6	7	8	9	10
11	12	13	14	15	16	17	18	19	20
21	22	23	24	25	26	27	28	29	30
31	32	33	34	35	36	37	38	39	40
41	42	43	44	45	46	47	48	49	50
51	52	53	54	55	56	57	58	59	60
61	62	63	64	65	66	67	68	69	70
71	72	73	74	75	76	77	78	79	80
81	82	83	84	85	86	87	88	89	90
91	92	93	94	95	96	97	98	99	100

72 + 3 = 75

Start at 72. Count on 3.
Say 73, 74, 75.

1. 37 + 2 = ________ **2.** 64 + 2 = ________

3. 50 + 3 = ________ **4.** 71 + 3 = ________

5. 17 + 2 = ________ **6.** 25 + 2 = ________

7. 39 + 3 = ________ **8.** 56 + 3 = ________

9. 77 + 3 = ________ **10.** 35 + 3 = ________

Name ____________________

Count On Tens to Add

Count on to add.
Use the hundred chart.

1	2	3	4	5	6	7	8	9	10
11	12	13	14	15	16	17	18	19	20
21	22	23	24	25	26	27	28	29	30
31	32	33	34	35	36	37	38	39	40
41	42	43	44	45	46	47	48	49	50
51	52	53	54	55	56	57	58	59	60
61	62	63	64	65	66	67	68	69	70
71	72	73	74	75	76	77	78	79	80
81	82	83	84	85	86	87	88	89	90
91	92	93	94	95	96	97	98	99	100

48 + 30 = 78

Start at 48. Count on 3 tens.
Say 58, 68, 78.

1. 24 + 20 = ______

2. 72 + 20 = ______

3. 85 + 10 = ______

4. 41 + 30 = ______

5. 36 + 30 = ______

6. 17 + 20 = ______

Continue the pattern.
Count by tens.
You can use the hundred chart.

7. 13, 23, 33, 43, ______, ______, ______

8. 38, 48, ______, ______, ______, ______

9. 29, 39, ______, ______, ______, ______, ______

Use with Grade 1, Chapter 28, Lesson 3, pages 517–518.

Name ____________________

Count Back Ones to Subtract

Count back to subtract.
Use the hundred chart.

1	2	3	4	5	6	7	8	9	10
11	12	13	14	15	16	17	18	19	20
21	22	23	24	25	26	27	28	29	30
31	32	33	34	35	36	37	38	39	40
41	42	43	44	45	46	47	48	49	50
51	52	53	54	55	56	57	58	59	60
61	62	63	64	65	66	67	68	69	70
71	72	73	74	75	76	77	78	79	80
81	82	83	84	85	86	87	88	89	90
91	92	93	94	95	96	97	98	99	100

65 − 3 = 62

Start at 65. Count back 3.
Say 64, 63, 62.

1. 58 − 2 = ______

2. 31 − 3 = ______

3. 47 − 3 = ______

4. 89 − 1 = ______

5. 19 − 2 = ______

6. 66 − 3 = ______

7. 42 − 3 = ______

8. 70 − 1 = ______

9. 78 − 3 = ______

10. 22 − 3 = ______

Name ______________________________

Count Back Tens to Subtract

Count back to subtract.
Use the hundred chart.

1	2	3	4	5	6	7	8	9	10
11	12	13	14	15	16	17	18	19	20
21	22	23	24	25	26	27	28	29	30
31	32	33	34	35	36	37	38	39	40
41	42	43	44	45	46	47	48	49	50
51	52	53	54	55	56	57	58	59	60
61	62	63	64	65	66	67	68	69	70
71	72	73	74	75	76	77	78	79	80
81	82	83	84	85	86	87	88	89	90
91	92	93	94	95	96	97	98	99	100

72 − 30 = 42

Start at 72. Count back 3 tens.
Say 62, 52, 42.

1. 86 − 20 = ______

2. 32 − 20 = ______

3. 71 − 10 = ______

4. 67 − 20 = ______

5. 22 − 20 = ______

6. 45 − 30 = ______

Continue the pattern.
Count back by tens.
You can use the hundred chart.

7. 81, 71, 61, 51, ______, ______, ______

8. 75, 65, ______, ______, ______, ______, ______

9. 68, 58, ______, ______, ______, ______, ______

Name ____________________

Estimate Sums and Differences

28-6
PRACTICE

Solve. Circle *Yes* or *No*.

Think first!
40 − 3 will be less than 40.

Think first!
64 + 20 will be greater than 64.

1. Will 90 + 3 be greater than 90?	Yes	No
2. Will 68 − 30 be less than 60?	Yes	No
3. Will 40 − 5 be less than 30?	Yes	No
4. Will 50 + 7 be greater than 60?	Yes	No
5. Will 50 − 20 be less than 20?	Yes	No
6. Will 42 + 30 be greater than 70?	Yes	No
7. Will 35 − 10 be less than 30?	Yes	No
8. Will 80 + 9 be greater than 90?	Yes	No
9. Will 61 + 20 be greater than 80?	Yes	No

Name ____________________

Problem Solving: Strategy

Guess and Check

Guess and check to solve.

1. Bob sees 2 kinds of flowers in the yard. He sees 38 flowers in all. Which two flowers does he see?

   ____

daisy	tulip	rose
20	3	18

2. Marlee sees 2 kinds of birds in the yard. She sees 21 birds in all. Which two birds does she see?

   ____

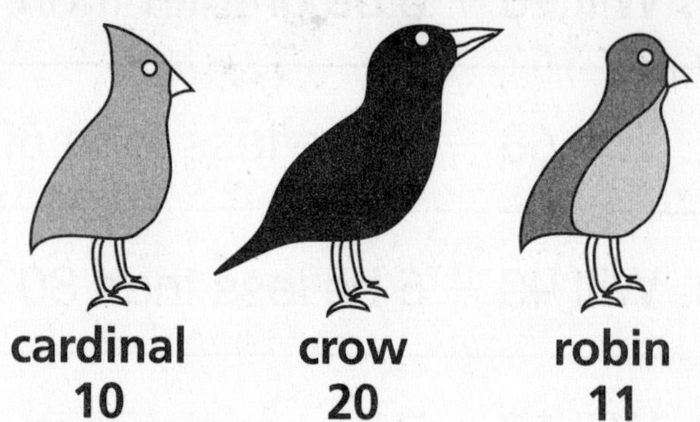

3. James sees 2 kinds of bugs in the yard. He sees 48 bugs in all. Which two bugs does he see?

   ____

grasshopper	ant	fly
30	20	28

4. Erika plants 2 packets of seeds. She plants 52 seeds in all. Which packets does she plant?

   ____

Summer Skills Refresher

Summer Skills

Orange Trees

In Florida there are a lot of orange trees. This is an orange tree. Florida's main crops are oranges and grapefruits.

1. One orange tree has 45 oranges. Show the number using tens and ones.

   ______ tens ______ ones

2. The second orange tree has 23 oranges. Show the number using tens and ones.

   ______ tens ______ ones

3. The third orange tree has 34 oranges. Show the number using tens and ones.

   ______ tens ______ ones

4. Which orange tree has the most oranges? Order the numbers from greatest to least.

   ______, ______, ______

Answers: 1. 4 tens 5 ones; 2. 2 tens 3 ones; 3. 3 tens 4 ones; 4. 45, 34, 23

The flowers on the orange tree are small and white. They smell very sweet.

5. On one tree there are 8 flowers on one branch, 4 flowers on another branch, and 3 flowers on a different branch. How many flowers are there?

______ + ______ + ______ = ______

6. Write the number of flowers using tens and ones.

______ tens ______ ones

7. On another tree, there are 7 flowers on one branch and 6 flowers on another branch. How many flowers are there on this tree?

______ + ______ = ______

8. Compare the number of flowers on both trees. Use >, <, or =.

______ ◯ ______

Answers: **5.** 8 + 4 + 3 = 15; **6.** 1 ten 5 ones; **7.** 7 + 6 = 13; **8.** $13 < 15$ or $15 > 13$

Summer Skills

The Manatee

A manatee is a large gray mammal that lives in the Florida waterways.

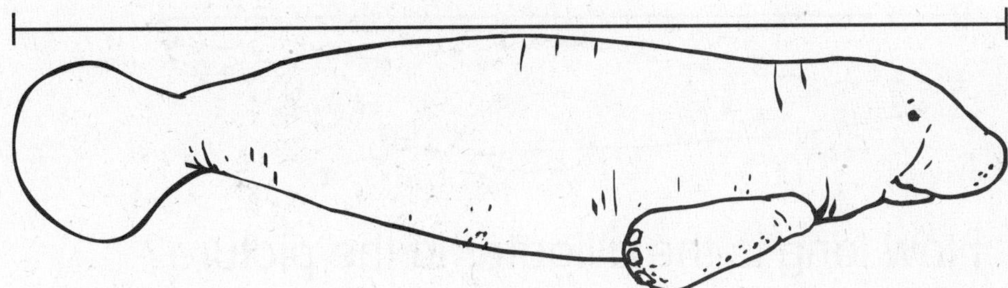

1. How long is the manatee in the picture? Estimate, then measure.

   Estimate: about ______ inches
   Measure: about ______ inches

2. The manatee has 2 forelimbs. How long is the limb in the picture?

   Estimate: about ______ inches
   Measure: about ______ inches

3. How many nails does the manatee have on one of the forelimbs? ______

4. If it has the same number of nails on the other forelimb, how many nails does the manatee have in all?

   ______ + ______ = ______

Answers: 1. 5 inches; 2. 1 inch; 3. 3; 4. 6

The alligator is Florida's reptile. It loves to sun itself on the river banks in the morning.

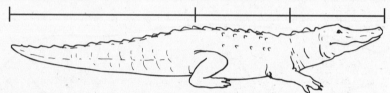

5. How long is the alligator in the picture? Estimate, then measure.

Estimate: about ______ inches
Measure: about ______ inches

6. The alligator's tail is half the length of the alligator. How long is the tail in the picture?

Estimate: about ______ inches
Measure: about ______ inches

7. How long is the head of the alligator in the picture?

Estimate: about ______ inches
Measure: about ______ inches

8. Compare the length of the tail and the head. Which is greater? Use > or <.

______ ◯ ______

Answers: 5. 4; 6. 2; 7. 1; 8. 1 < 2

Summer Skills

Shapes and Figures

There are a lot of different hotels in Florida.

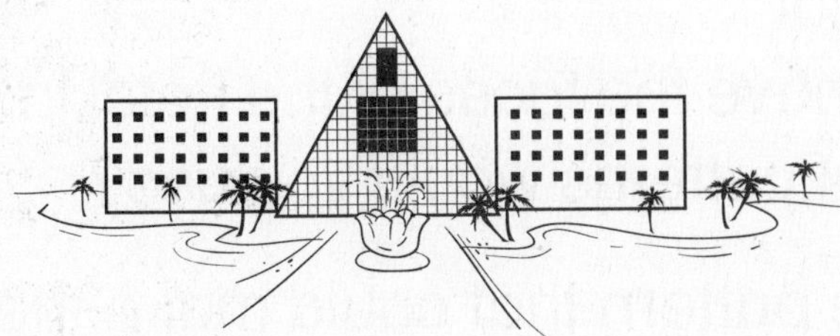

1. Draw the shapes of the windows. Then write the name of the shapes.

____

2. What are the 2-dimensional shapes that make up the two sides of the hotel? Draw the figure and write the name of the shapes.

____

3. What is the 2-dimensional shape in the middle of the hotel? Draw the figure and write the name of the shapes.

____

Answers: 1. square; 2. rectangle; 3. triangle

If you made the 3-dimensional figures that have the faces of the hotel in Florida, what patterns would you use?

4. Circle the pattern that could make the 3-dimensional figure that makes each side of the hotel.

5. Circle the pattern that could make the middle figure in the hotel.

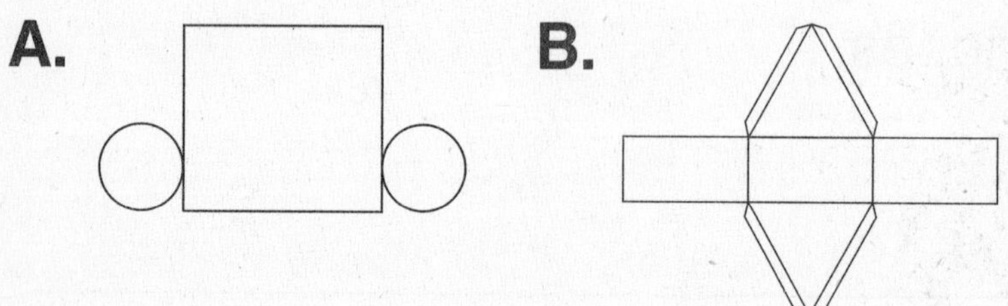

6. What figure could the pattern below make? Draw a picture and write the name.

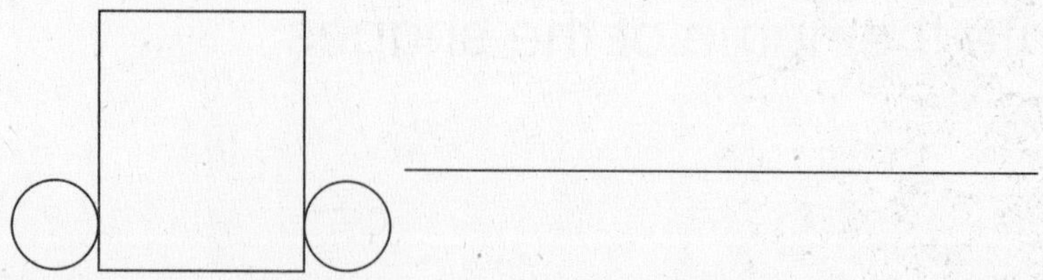

Answers: **4.** A, the rectangular prism; **5.** B, triangular prism; **6.** cylinder

Summer Skills

What Is Missing?

Erik's family went to a parrot jungle on the weekend. There are parrots, cockatoos, and macaws.

1. Erik saw 12 birds. 5 were parrots. How many birds did he see that were not parrots?

_____ + 5 = 12

2. Erik's brother, Marc, likes the macaws. He saw 14 birds. 9 of the birds are macaws. How many other birds did he see?

9 + _____ = 14

3. Erik's mom likes the cockatoos and parrots. She saw 16 birds. 8 of the birds are cockatoos. How many are parrots?

_____ + 8 = 16

4. Erik's dad saw 15 birds. He saw 6 cockatoos. The rest of the birds are macaws. How many macaws did he see?

6 + _____ = 15

Answers: 1. 7; 2. 5; 3. 8; 4. 9

After lunch at the Parrot Café, they went to the petting zoo.

5. Erik counted 11 baby birds. 6 are ducklings. How many are chicks?

_____ + 6 = 11

6. Marc is counting the lambs. He counted 12 in all. How did he count?

_____, _____, 6, _____, _____, 12

counting by _____

7. Erik's mom petted the piglets. She petted 8 piglets. What is that the double of?

_____ + _____ = 8

8 is the double of _____

8. Erik's dad used mental math to add all the animals his family had played with. How did he do it?

_____ + _____ = 20

20 + _____ = _____

Answers: **5.** 5; **6.** 2, 4, 8, 10, counting by 2s; **7.** 4 + 4 = 8, 8 is the double of 4; **8.** 12 + 8 = 20, 20 + 11 = 31

Summer Skills

The Sunshine State

Most of Florida enjoys sunshine almost every day of the year. Use the graph to answer the questions.

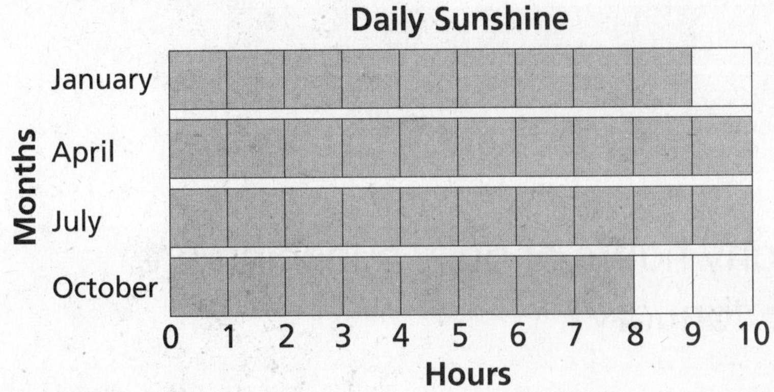

1. Which month has the fewest hours of daily sunshine?

____

2. How many hours of sunshine is that?

______ hours

3. Which months have the most hours of daily sunshine?

______________, ______________

4. How many hours of sunshine is that?

______ hours

Answers: **1.** October; **2.** 8 hours; **3.** April and July; **4.** 10 hours

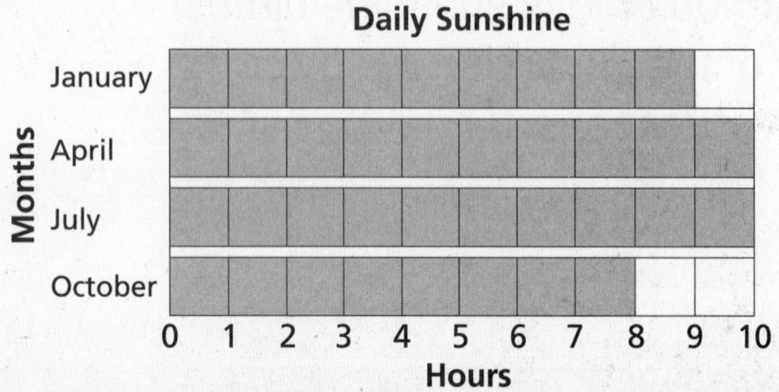

5. How many hours of daily sunshine are there in January?

_____ hours

6. What is the range?

____

7. What is the mode of the numbers?

____

8. If you like sunshine, when would you visit Florida?

____

Answers: **5.** 9 hours; **6.** 2; **7.** 10; **8.** April or July